To Herman

Aliens Above, Always

A UFO Report

With best wishes

John Magor

Dedication

This study is dedicated to the developing association between us and our UFO visitors which, though still barely recognized, may set us on an utterly new course in history. While some observers have said these visitors are a passing fancy, signs persist they will be here always.

Herman J. Aaftink
Calgary, Alberta,
2 June, 1987

Aliens Above, Always

A UFO Report

John Magor

hancock house

ISBN 0-88839-969-3

Catalog in Publication Data

Magor, John, 1915-
Aliens above

Includes index.
ISBN 0-88839-969-3

1. Unidentified flying objects — Sightings and encounters.
I. Title.
TL789.M33 001.9'42 C83-091206-7

Edited by Diana Ottosen
Typeset by Elizabeth Grant in Times Roman
on an AM Varityper Comp/Edit
Production, Layout & Cover Design by Eva Raidl
Printed in Canada by Friesen Printers

hancock house

Hancock House Publishers Ltd.
19313 Zero Ave., Surrey, B.C. Canada V3S 5J9

Hancock House Publishers Inc.
1431 Harrison Avenue, Blaine, WA, U.S.A. 98230

Table of Contents

Photo Sections

COVER PHOTO

This pair of UFOs was seen dancing in the sunlight over Yungay, Peru, in March, 1967. (ICUFON Archives)

Preface
by
Brinsley Le Poer Trench
Earl of Clancarty

This is a fascinating book. The author covers a lot of UFO operations that many people interested in the subject may not have heard about. For instance, the UFO device referred to as a "bell-jar", which, to judge from incidents described here, may well be an overlooked key to many of the secrets of the sky people.

Then there is the flying instrument apparently used for marking cattle, as well as what may be portable bolts of energy, to mention just two more remarkable items. Extraordinary, too, is the eyewitness account of an animal taken by light-ray aboard a UFO.

It is obvious that the aliens are using some advanced technology and abilities far beyond our own. Indeed, they often use what we would term paranormal powers which are quite normal to them.

Now, from where do the aliens emanate? Personally, I am more inclined to the extra-terrestrial proposition than that the visitors come from some other dimension. I think that the latter theory has come about because the visitors have these paranormal powers, and this caused them to be called "gods" in ancient times.

I consider these visitors to be from very advanced civilizations in our galaxy. It seems that they use in their craft a kind of free electro-magnetic energy which enables them to overcome gravity and get anywhere very quickly indeed.

Furthermore, I go along with the concept that the aliens have been coming here for thousands of years, probably from many planets. So, different coloured races may have come here, and either mutated with primitive earth man, or settled here. This could account for the five basic colours here on Earth: white, black, brown, red and yellow. All countries everywhere have their legends of aliens flying in. China has "Dragons" flying over that country. Japan has "Dogons". The Hopi have their tales of alien visitors in North America. And so it goes on all round the world.

I agree with Sir Fred Hoyle's suggestion that there may be thousands of advanced civilizations in our galaxy communicating with each other, and that we are as unaware of it as the pygmies in the African forests are unaware of the radio messages flashing round the earth.

In short, I think that we are a space colony, and the aliens keep coming back to check on our progress. I reckon that they must be a little worried about us right now with our increasing violence, nuclear weapons and pollution.

At one time we may well have all had what we now call paranormal powers. There are still a few people on this planet who have some of these abilities. It is more than likely that mankind lost these powers during the catastrophes and ice ages that have taken place during our long history.

There is a lot of most interesting material in this book, and a good deal to be learnt from it. So, without further ado I will now hand over to the author, John Magor, for a splendid read and much food for thought.

London, England

Already known for his abundant literary work on UFOs, Lord Clancarty made another important contribution in 1979 when he initiated a major debate on the subject in the House of Lords, the first-ever such debate by any legislative assembly anywhere. He followed this by bringing together an all-party UFO study group in the Upper Chamber which invites talks by ufologists, and others with related interests, from various parts of the world.

Chapter One: Our Watchful Visitors

Brian James's concept of a sunlit craft leaving the moon.

Those who find it hard to believe that Earth is visited by intelligent creatures from space will generally contend the distances between this and other worlds are too great. While they may concede there must be other civilizations out there in the vastness of the universe, and may further concede that UFO action in our skies seems to be a real mystery, they prefer not to think there is any connection between the two. For how could traffic from space match the frequency with which UFOs are observed?

What apparently is missing from this line of thought is any supposition that many, or even most, of these visitors may originate in our own solar system. If their intelligence is as superior to ours as it seems to be, is it not possible they have lived and multiplied here no matter how distant their ancestral homes?

There is a clue in the behaviour of UFOs that suggests this is indeed the case. The clue lies in their evident awareness of earthly affairs, an awareness that seems possible only by close and continuing observation.

Since the aim of this book is to add to the substance pictured for our strange visitors — in other words, to discern their purpose — I think a listing of incidents pointing to that awareness is appropriate for our first chapter. However, to do justice to the sense of organization that seems to lie behind these cases, we need a story tying them together. So let's recreate certain events as best we can, basing our account on verified UFO material. For the local setting, let's link our story to a few anomalies observed on the moon and visualize ourselves in a scenario that goes something like this:

We are members of an alien colony living in chambers deep inside the moon. Although years have passed since astronauts made their brief visits to the surface, our colony has observed the increasing number of space vehicles dispatched from Earth and realizes that eventually the visits will be resumed. Perhaps the next part of the program will be more aggressive as humans start competing for a place up here.

To moderate their actions by making them realize they will have company, one of us with a knowledge of earthly language has been assigned to draft an account of our presence without, of course, revealing how the moon's interior can be entered. He will also explain what steps we have in mind for maintaining contact. Later our governing council will decide to whom on Earth the message shall be addressed and how its source will be identified.

Now, our colleague writes:

Eons ago when your sun was to us a faint new star, our ancestors like so many others elsewhere, embarked into space from their distant home to follow a colonist's way of life in the universe. Where they encountered other civilizations their mission was not to interfere but to establish outposts from which they could keep their mother planet informed about the affairs of these developing worlds.

Countless lifetimes later and now light years out in space, our parents and the companions who were left continued this tradition by undertaking a journey that brought them to your solar system. Here they singled out Earth's moon to raise their families for the next voyage to the stars. They chose it because it is an old flagship from another system, whose hull is the quarried husk of a once natural satellite. Although abandoned because of damage suffered on its flight through the asteroid belt, which was then a violent new danger zone after the planetary explosion between Jupiter and Mars, the moon is now known to colonists as a reasonably safe refuge from the hazards of interstellar travel.

There is ample living space inside the moon where I write. Indeed there is so much room that when, as we have learned, your astronauts tested the structure of this vehicle by slamming a discarded module against it, they were surprised that it vibrated as if ringing like a bell. We, of course, hardly noticed that little experiment. An entry in the ship's log reports that the lunar walls started to tremble when they were hit on that flight past Jupiter, and they have never stopped.

Although we colonists can now move freely about the surface and sometimes use lights that no doubt puzzle your astronomers, we also use mobile observation posts we can operate from inside the moon. These must be even a greater mystery to the astronomers. Obviously they will wonder what is happening when little domes appear and then vanish on the surface of this "lifeless" satellite.

But perhaps best of all for us is that the machinery which propelled the moon through space and established its beautiful circular orbit is still in good working order, because we have repaired and kept it that way. In fact, an essential part of it continues to keep the same lunar side facing the earth millenium after millenium, just as the engineers who converted the moon from a natural satellite into a spaceship had planned. This is an important factor in making the vessel more secure. With several worlds to visit, it meant the crew would know where their defence must be concentrated in case of an attack from the one being surveyed.

Although an attack from Earth is quite improbable at present, remember that at some future time the moon will become a military prize of great strategic value sought by your leading nations. That in turn will affect us as the tenants and is one of the reasons we must now make ourselves known while continuing to keep a sharp watch on what you are doing. I will return to that point in a moment.

Now, as your observers have recorded, the moon oscillates slightly on its vertical axis, briefly revealing edges of the far side. This is a result of the moon's natural tendency to rotate in orbit working against the artificial action of its gyroscopic controls. The faulty movement reflects the outmoded design of its equipment. However, thinking about the great age of that equipment and the work necessary to maintain it, we are gratified the problem is so slight.

It is not simply that we must keep the antiquated machinery going (a single "saucer" of modern design can produce as much power as the moon's entire engine room!) but we must constantly rebuild it. To do this we mine and fabricate the material for all the parts, just as any civilization must do.

Our best source of mineral supply, now supplemented to an increasing extent by ore from the asteroids, is on the far side where the impact craters are much more numerous. This means that a great deal of ore-bearing rock is exposed, making it easier to mine.

Here I should explain that in flight the back of the moon served as the front, if you follow me. In other words, it was the bow of the ship as it went sailing through space, and naturally it took most of the brunt of collisions with space debris.

In their wisdom, those who restructured the moon to visit ports of the universe planned it just that way. They saw that a rotating ship, though providing better balance and helping to solve the gravity problem for those inside, would not be as serviceable on arrival, having been battered on all sides. Therefore most of the power source was located in one section in the stern and, for each trip, the automatic pilot would be reset to maintain a fixed heading until the flight crew disengaged it. Thus on slowly approaching their intended distance from Earth, and despite that awful damage experienced after passing the Jovian orbit, they released the controls and rotated the craft until the stern faced forward. So the side of the moon you see from Earth, with its ancient beds of spent fuel known as "maria", is relatively smooth.

As I have said, keeping the same side facing the most likely source of danger was part of the original plan for this vessel. To help it carry out its duties as a flagship touring the universe, it was always assured of a clear view of its ports of call. Any direct flight and possible landing from one of those ports would be spotted immediately.

Now I come to our purpose in writing. Although we have had no visits from you lately, we are fully aware that the space age for your people is at hand and that, by the same token, so is the end of our quiet life as a colony. Although we have the option of removing the moon from its orbit and finding other facilities to replace those we now enjoy as your neighbor, our council has decided the wiser course is first to confirm our presence (which to some extent you have already guessed) and then observe your reaction. Since any message from space will be suspect, including this one, we believe contact can be most effectively established by listing recent earthly events which alarmed us and by reminding you how we showed ourselves on those occasions.

Here you must understand that as the visual method of communication has proved the most workable for our colonists in other solar systems, removing as it does all doubt that they are the senders, we too will maintain our contact with you by that means. Therefore after transmission of this introductory message, do not look for radio signals from us. Any further messages will be sent by our displays in the sky. We will make them increasingly specific once we are sure you are paying attention.

Now let's review what we have shown you recently. We will start with the war in the South Atlantic and go back a few years from that point.

When we observed in the Fall of 1981 that Argentina was making plans to invade the Falkland Islands we realized, of course, that the target area was of little significance itself. However, we were uneasy about the dangers since the British and perhaps the Americans would be quickly involved. Therefore we contacted other nearby forces in the solar system and together arranged a display that would be widely seen over Argentina and provoke controversy, thus increasing its effect.

Please understand that when we see a crisis developing on Earth we stage these displays not with any belligerent intent but to let you know we are keeping an eye on your affairs. In this case we chose a time when two things would happen simultaneously in or over the same geographical area that concerned the danger of global war. First and most immediate, of course, was the gathering of Argentine forces. Second was another sort of warlike activity by another country that was already in progress literally far above the Argentine preparations. I refer to

Russia's space experiments in sending aloft its early warning Kosmos satellites.

Having noticed that these satellites, launched from Siberia and circling the world, travelled high above Argentina where their luminous "propellent dumps" startled the crowds below,[1] we decided to take part in the spectacle. Our purpose was not to frighten but to create doubt that all of these displays were of human origin.

Accordingly while Argentina went ahead with its invasion plans, we made our own preparations for the next launching of a Soviet rocket. Our lookouts saw that happen the night of Oct. 15, 1981. Then, as the satellite approached South America as a point of light at great altitude, our brilliantly lit device rose quickly into the air. At no time, however, did we let it be seen near the satellite's glowing cloud-like "dump" which was discharged offshore and continued along on its own. Two enormous unexplained lights together in the sky could have caused panic which, to repeat, was not our intention. Instead, by flying separately at low altitude and by moving erratically so that at one point our device almost collided with two airliners,[2] we caused confusion, which was our purpose.

Because the pilots talked openly of a UFO flying about at their level, and one of them even reported seeing a metallic ball,[3] — it was part of our lighting equipment — it will be hard for sceptics to convince everyone we were not there that night as Argentina prepared for war.

Going back two more years we saw another instance when a threat of war developed. The scene that time was Poland where workers were organizing in revolt against the hard conditions imposed on them. With the danger of Russian action hanging over the country, we decided to make a showy appearance over an area where we were sure to be seen. Accordingly on the night of August 20, 1979, we had four of our craft, brilliantly lit, fly slowly in triangular formation over the central airport at Lodz. Then to emphasize our presence, we had a group of ten fly over Warsaw, the capital.[4]

While we make no claim to being alone in foreseeing the troubles in Poland, you may find our timing of interest for it was exactly one year to the day after our performance that the Polish leader Lech Walesa proclaimed formation of the Solidarity movement. Although we hold no brief for or against the politics of humanity and are skilled observers rather than prophets, we abhor oppression and are pleased we anticipated events accurately enough to accord these anniversary honors ahead of time to the Polish people for their brave gesture.

(1) OBERG, JAMES, in *Frontiers of Science*, vol. 4, no. 3 (July-August 1982).

(2) *Tual Cual* magazine, Buenos Aires, Argentina, 6 November 1981. *Tribuna Rosario*, Rosario, Argentina, 2 November 1981.

(3) *Tual Cual.*

(4) *Gazet van Anterpen*, Antwerp, Belgium, 25 August 1979.

(To this we should add that occasionally our crews make landings to assess a situation literally at ground level. In the Polish case we had a crew of four operate near Lublin in the vicinity of Lodz. When they were seen by a farmer on May 10, 1978, they took him for examination, releasing him after a few minutes. Later one of our crew was spotted by reporters at the scene.[5]

(A crew of four also operated at ground level in the Argentine case. Two months following our display there its members were seen by townspeople after landing their craft in a park in Rosario, in Santa Fe province.[6] Because of that exposure, they left quickly.)

Now let's go back a bit further and to yet another part of the world. If you remember the international events of the time, you will recall this was a period when negotiations with oil-producing countries in the Middle East were critically difficult. There was even talk of using force if necessary to keep the oil traffic moving.

Watching all this with concern, we decided to reveal ourselves in a way that would dramatize the problem and show that we knew precisely where the danger was centered. Thus on the night of October 19, 1978, one of our crews staged a boldly visible landing at an oil-pumping station in Kuwait at the head of the Persian Gulf. While a group of frightened workers looked on, our crew emphasized its point by stopping all pumps at the station.[7]

Visually as well as we could, we had told your people: "We have seen that this place, where oil is shipped, is your world's danger point. Take note that we observe."

As our craft rose in the air to leave, we allowed pumping operations to start again.

In a moment we shall consider other events of that year, 1978, as it was one of outstanding importance in our efforts to become known to you. But first let's go back just another two years to illustrate again the devices we use to communicate visually.

You have seen that when an entire nation appears in danger of taking a disastrous step, we may offer a brilliant single display visible to a large part of the country concerned, as in the case of Argentina. When some individual achievement of great but alarming importance is involved we think a flight of luminous craft moving in formation, particularly over a major airfield as in Poland's case, is a suitable medium.

To focus attention on a certain locality, or even on a certain activity, which could become the central feature of far-reaching problems, we believe a dramatic but localized performance of the sort presented at Kuwait is most appropriate.

(5) IWANEC, MAREK S., in *Journal UFO*, vol. 2, no. 3 (December 1980) also letter from DEUSSING, W.F., Ottawa, Ont.

(6) *La Razon*, Buenos Aires, Argentina, 11 December 1981.

(7) The APRO bulletin, vol. 27, no. 7 (January 1979).

Brian James

Occasionally, however, we are faced with a situation which has elements of all of these examples. It may involve a country with serious internal issues at stake, and perhaps a leadership problem, which though largely local in nature could flare into a crisis with international impact. How can we show to humanity that we have observed all the broad dangers, yet arrange a display in keeping with the narrow physical limits of the disturbance?

As you may now realize, Iran gave us exactly that kind of test a few years ago, and you may remember how we replied. Since a leadership question was involved, as well as all the other supplementary issues, we chose a major airport for our setting as in the Polish case. Thus on the night of September 19, 1976, we staged a display close to the large airfield outside Tehran. But on that occasion our purpose was to pinpoint a quarrel rather than to pay respects. Therefore, instead of a ceremonial flight, we introduced a single brightly lit craft which flew past in a manner that invited trouble.

The challenge was met by the Iranian air force which sent up two fighters to intercept. When the first fighter attacked, our "UFO" stayed in position and knocked out the craft's instruments and radio, causing it to withdraw. The second fighter received the same welcome, except its weapons-control was also knocked out to save us the trouble of causing further damage. Knowing the pilots might have trouble explaining what happened, we dropped a flare at the scene to confirm we were there.[8]

Two years later Iran reached a state of civic upheaval, as we had expected. This led to banishment of the Shah and then to seizure of the U.S. Embassy. It was a period of frightening events and, even though we could foresee more bad times for the Middle East, we had some satisfaction in knowing the drama we staged near the Mahrabad airport had accurately pictured our fears. We did not relish causing trouble but how else could we impress you with our presence and our awareness of strife ahead?

Before dismissing all that as just another unusual UFO incident in case you doubt the authenticity of this message, please remember that to some degree our strategy worked. While it may not have changed the course of events in the Middle East, for the present anyway, your officials at least paused long enough to record our presence. As you know, the U.S. Pentagon received a detailed report of the whole affair.

Now, from the Iranian episode, let's turn to what in our opinion is the most memorable performance we have ever staged for the benefit of mankind — next, of course, to the Star of Bethelehem. This is what you would call the "flap" of 1978. Never before have we offered such a panorama of sights in the sky and earthly encounters.

Since our plan was to gain your attention as fully as possible, we had the participation of our underwater stations on Earth and our bases on Mars. Also we had help from moons of the outer planets where we have assembly plants operated mainly by robots. (Even though you know us as super-beings, there are limits to what we can endure!)

As our stagings were well reported by your news media, here are some in that year you should remember, in addition to the incident in Kuwait.

One of our largest displays was presented over Italy just before dawn on Sept. 14. In that instance we had a craft, which your reports described as being "the size of an ocean liner", fly from Sicily to a point beyond Rome. It was brilliantly lit and equipped with shining wings.[9]

As we were specially concerned with affairs in the Mediterranean area, for reasons I will explain in a moment, we followed this two days later with three successive visits by an illuminated craft flying directly over Rome. Earlier that year we had put on a three-hour show with a squadron of "UFOs" performing aerobatics over the Pyrenees reaching the western end of the Mediterranean.[10]

(8) *International UFO Reporter*, vol. 3 no. 1 (January 1978) & others.
(9) *Review Record*, Miami, Florida, 19 December 1978, & others.
(10) *Canadian UFO Report*, vol. 5, no. 1 (Winter 1978-79).

Unfortunately, to compete with the sensational news your people like so well, we must occasionally resort to alarming tactics to remind you that we exist. In 1978 we did this on several conspicuous occasions on a worldwide scale. In one of these cases we had a particularly large craft hover over a moving automobile and take it aboard. That was on a September day in Chile. A father and his son were in the car and we let them have a good look at our five large crewmen. No doubt we scared them, too, but we let them go safely after making sure there were witnesses to confirm what happened.[11]

Then there was the time in February off the coast of Argentina when we put on a show for a group of fishermen. It was not quite dawn as our craft descended and one of our tallest members in a shining silver suit climbed down a ladder hanging near the boats. He waved at them before climbing back again, satisfied he had given them something to remember.[12]

But for our climax that year we chose someone who was so interested in our activities that we knew we could safely take him with us and thus in a personal way commemorate that eventful and significant period. He was Frederick Valentich, an experienced Australian aircraft pilot whom we had monitored carefully before his take-off on Oct. 21 on a flight over water near Melbourne.

As our craft chose the right moment to approach, we could hear him on his wireless, saying, as quoted in your world media, "It seems to be playing some sort of game. . . it is a long shape. . .cannot identify more than that. . .coming for me right now. . .it seems to be stationary. . .I'm orbiting and the thing is orbiting on top of me. . .it has a green light and sort of metallic light on the outside. . . ."

Then we took him, careful not to leave a trace of his craft to suggest he had crashed, and our choice was a good one. Like many others of your people, Fred is living happily with us now and still studying all those things he was curious about.

Now you may ask, what was the purpose of all that? Why did we choose 1978 to make our extraordinary effort to impress you with our presence?

Remember I said we were concerned with affairs in the Mediterranean area, which to us means primarily affairs of the Holy Land. For us, as it does for you, that part of Earth symbolizes more than any other the religious and cultural cradle of mankind, and in relationship to its problems we offer the most visible clues to our presence.

In 1947, at the start of a new cycle of appearances, we gave you a good example of that when we made a well-publicized flight over mountains in the State of Washington. Except that it was in an area uncluttered by other distracting events and offered a dramatically good view, the place was not important but the timing was, for that was in the

(11) Ibid.

(12) Ibid.

Liking the cloud effect at the top of a mountain near Kelsey Bay, Vancouver Island, Hannah McRoberts of nearby Campbell River took this shot in October, 1981. Flying disc at right was seen only when film was developed. Experts at the Vancouver Planetarium found no flaw in the photo. What looks like figure inside dome is visible when original print is seen under a magnifying glass!

same period that the State of Israel was founded and the Arab-Israeli wars began.

That introductory flight was followed by our mass appearances in 1952, 1956, 1967, and 1972, all timed to coincide with major crises in the Middle East centered on the Holy Land.

Then, coming to the memorable year of 1978, we had seen not only the dangers in Iran and the shape of more disasters to come for people of the Mediterranean, but this time a new threat appeared as Russia prepared for its march to the south. Once more we showed ourselves in strength in the manner I have explained.

That was not all, however. If you have followed reports about us carefully enough, you will remember that a few days after the year's end we staged our finale. On that occasion three of our craft, illuminated in many colours, passed in broad daylight over the Mount of Olives in Jerusalem.[13]

(13) Ibid.

Take note of that symbol of Trinity and the sacred place we chose for it. Surely you will not doubt now that you have company overhead, informed and watchful!

While the story linking these episodes is fictional, of course, the episodes themselves are not. As listed in the reference notes, all these episodes are taken from sightings on record. If the picture was presented well enough, we can see with a little imagination that our strange visitors act with a sense of purpose.

So, in the following chapters, let's dispense with the storyteller and move directly into several broad areas of UFO activity where our visitors further reveal they have something in mind. We may not understand all about that "something" but, again if the picture is presented well enough, we will see it is there.

With wide use from other sources, my material is taken for the main part from records accumulated over ten years of publishing *Canadian UFO Report*, referred to here as CUFOR or sometimes as "we". While my base of operations is Vancouver Island, facing out to the Pacific on the coast of British Columbia, the scope of this effort is international, in keeping with the nature of the subject.

As it appears that UFOs generally approach their earthly problems with care and respect, I have attempted to return the compliment by offering their performance in some sort of order. As we have studied in this chapter, there is what I consider the main area of connection between these aliens and ourselves. Here we see them almost pointedly keeping watch on our planet and human affairs. Their motive for doing so seems partly for our benefit, such as in those cases just reviewed where they try to make us aware of their presence, and partly to meet some requirement of their own such as maintaining navigational aids.

Next we will examine an extraordinary device, here referred to as the "bell-jar", which seems to be an essential instrument in much of the UFO program. While its main function may be the awesome one of creating various forms of life and causing them to disappear as well, it may also provide a view of areas beyond known dimensions. If it does, the bell-jar becomes an important part of the answer to a vital question in ufology — are these visitors extraterrestrial (ET) of ultraterrestrial (UT)?

At the same time it may help explain how those seemingly tough colonists, building and operating nuts-and-bolts equipment in our planetary system, if this is where some of them live, are also capable of the ghost-like activities so often observed on the UFO scene.

In view of its obviously great importance, different aspects of this device are studied in three consecutive chapters. One of them concerns the remarkable experience of a man who reported he was made captive inside a bell-jar, an incident that illustrates some of the device's extraordinary properties.

Branching from the bell-jar theme, we look at another less mysterious tool basic to UFO needs. It is electrical power, and in two chapters we see how eagerly the aliens hunt for it and how they use it, in a form unknown to us, in work on their navigational aids.

Animal mutilation, much in the news from time to time and unconvincingly blamed on cults and natural predators, appears to provide a forum of its own for UFO activity. In the final chapters we examine that scene after reviewing an incident in 1972 that may have prepared the way for all that followed. It seems reasonable to guess that this sequence of events could lead to a major encounter with mankind, so boldly have the aliens shown themselves.

What Is Going On Up There?

Formations like roads and bridges sometimes border smooth depressed areas on the moon known as *marias* for seas just as highways follow our shorelines. Arrows point out such features in this view of Mare Crisium.

Our scenario in chapter one picturing the moon as an old spaceship now serving as a hostel for space colonists may not be entirely fanciful. Despite the opinions of so-called realists, puzzling NASA photographs as well as observations from Earth raise a question not fully answered. What really is going on up there?

Following up on George Leonard's *Somebody Else Is On The Moon*, contributor Graham Conway obtained a batch of NASA photos that seem part of the riddle. Later we received more fine examples from Jim Safran of Van Nuys, California, plus a few unusual shots of Mars. Those shown here are selected from that supply, which was supplemented by some already in CUFOR files. Three or four of these shots have previously been well publicized but none thoroughly explained.

Although tiny details are always lost in reproduction, just enough comes through here with the help of a strong magnifying glass to confirm a mystery does exist.

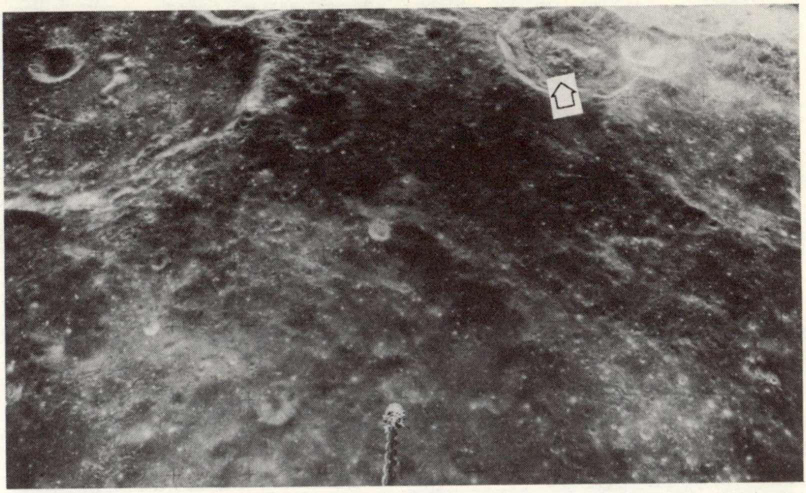

Far side of the moon with its many craters may be a good mining area for *colonists*. Arrow points to T-shaped feature in crater which looks like a piece of equipment. Same feature is seen elsewhere on the moon.

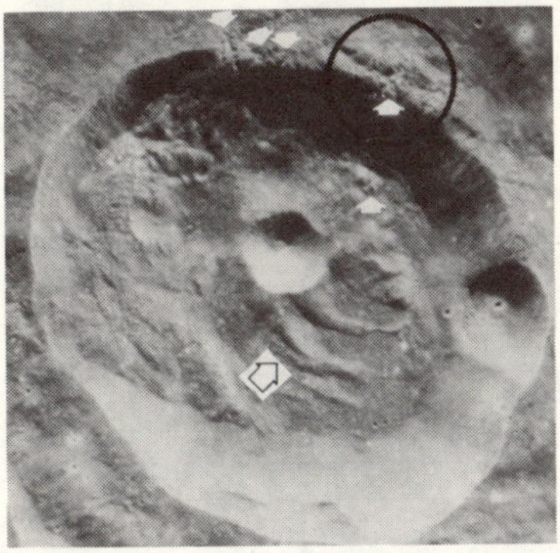

Are strip-miners at work in this photo of Tycho Crater? Ridges inside the rim at bottom make it appear so. The second arrow farther north points to a domed complex that may be part of the operation. Small bumps on the rim inside the circle may serve as vehicles while arrows at left indicate gouges which suggest more digging. Among moon watchers, Tycho has long had a name for its *Lunar Transient Phenomena* or LTP.

Area of Mare Orientale, Mare Veris and Rook Mountains seems alive with roads (not to be confused here with scanning lines). While some obviously are surface cracks, arrows point to sections of whitish lines that appear to cross cracks.

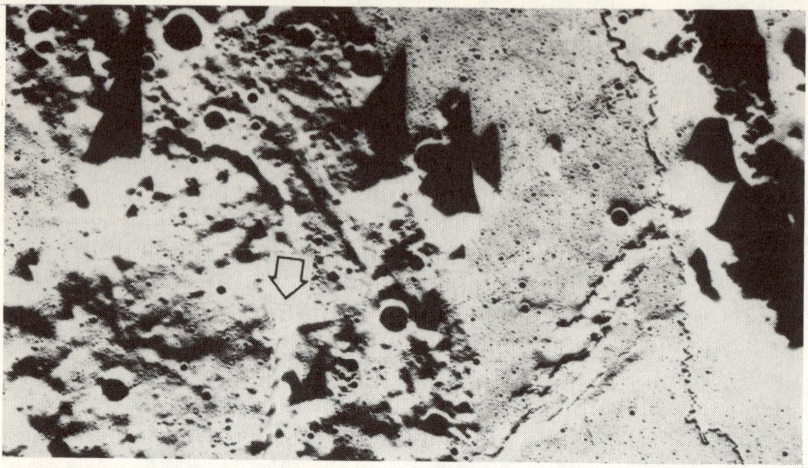

Like something in witch country, a cluster of points in this starkly lit section of lunar surface casts long sharp shadows. *Pyramids* like these raise questions of alien design. Arrow at bottom points to five *pylons* in straight line.

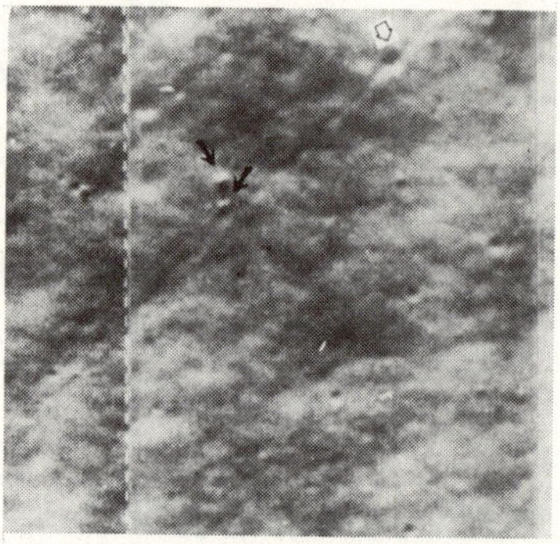

Objects rolling in opposite directions? Apparently they are in this action shot of the far side of moon. Judging from the track, the large object at lower left started at upper right and went through a hole (white arrow) en route. Leaving fainter parallel tracks, the two other objects seem to be going the other way.

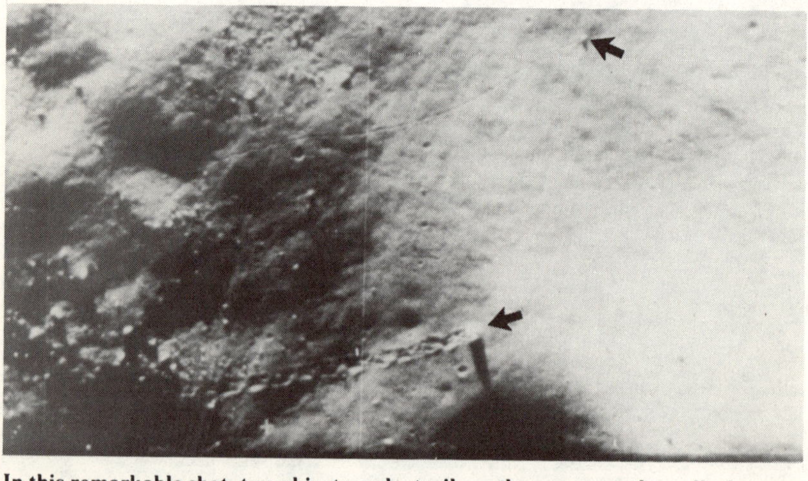

In this remarkable shot, two objects make trails as they move up the wall of crater Vitello. Since the larger one leaves tread-like marks, it appears to be climbing under its own power. The pointed shadow confirms it is not simply a round boulder that is somehow rolling uphill. The trail of the smaller object shows it has done a difficult trick climbing from the hole at left.

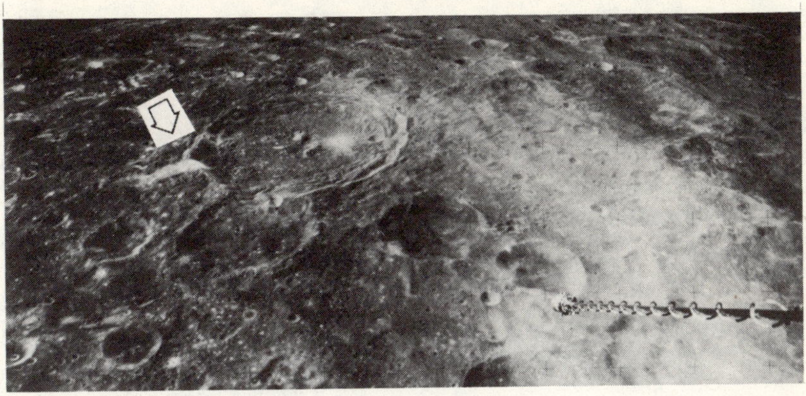

On rugged far side of the moon, unexplained white *gusher* seems to flow over the rim of Lobachevsky Crater. The boom of orbiting module protrudes at right.

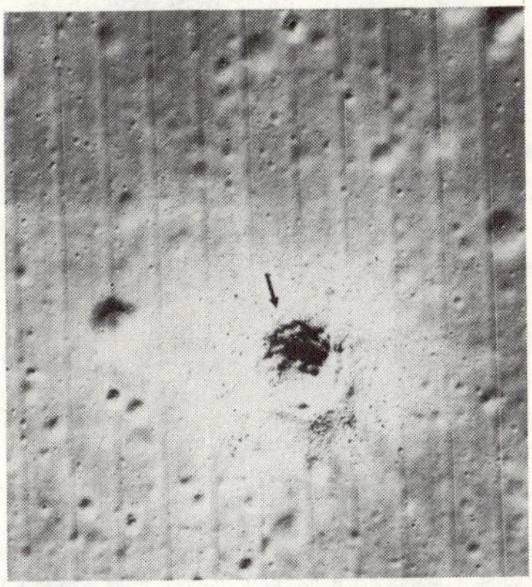

Unlike nearby craters, this small one in Oceanus Procellarum reflects field of light and, surrounded with debris, is jagged as if digging is underway. Two bumps that may be equipment are inside the top part of rim.

Mars has its own riddles as does the moon. In this 1976 shot from the Viking Lander, letter B on rock is circled at left. It was said to be due to lighting, but to the right is a white nozzle which appears to be squirting something on the ground where it bubbles or piles up. Below that is an arrow which appears to be carved on rock. Evidently, martians like grouping their artifacts as the next photo suggests.

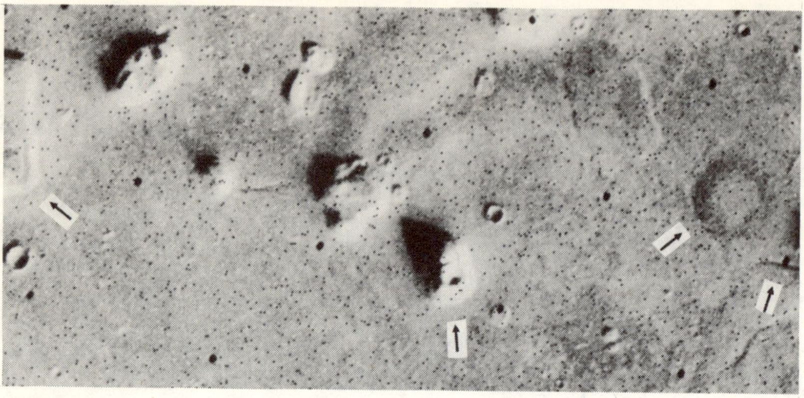

While NASA explains this now-famous face on Mars is "formed by shadows on eroded landforms," some analysts contend aliens may have designed rock with those shadows in mind. Other factors favor this idea of artificial origin. At far right is a form with an unusual right-angle edge while at center-left there is an indentation as if the ground had settled around a circular structure. To the left there is a T-shape leading to a sharp-edged rock which is topped with a squarish black mark or opening. Left of this opening triangular shadow shows a pyramid-like rock. The entire area suggests this is a special place. The width of the head is one mile.

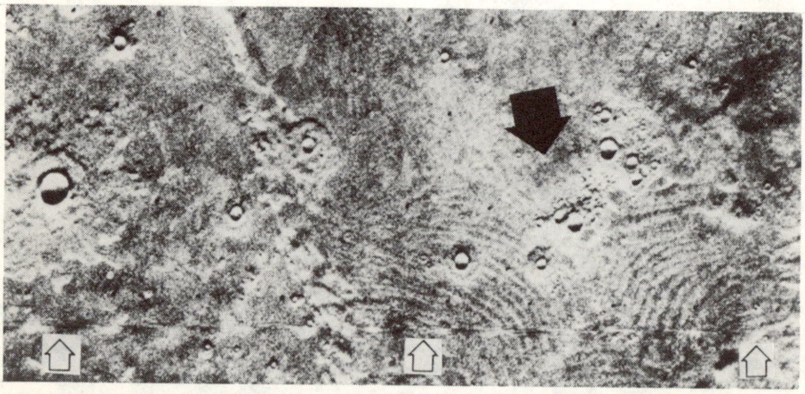

An area resembling plowed fields, which the black arrow indicates, appears in this photo of the Cydonia region of Mars and measures about 40 miles square. NASA explains that the furrows are the result of wind erosion, but something else looks strange. The white arrows point at what looks like a trench cutting through the landscape. Though not clear in reproduction, the trench contains at least two parallel tracks that seem part of the terrain, unlike the processing lines in other photos.

Chapter Two: That All-Purpose Bell-Jar

Like the super-magicians they seem to be, our UFO visitors apparently have an endless assortment of devices to create their mystifying effects. In addition to the major displays we considered in chapter one, the aliens sometimes stage much smaller performances with such theatrical impact, it seems their only purpose is to impress a few spectators with their showmanship.

That, of course, is unlikely. It is more probable they are simply doing what comes naturally and what in one way or another has a specific purpose. Even so, many of their tricks do have a stagey look and to some extent could be duplicated by any clever vaudeville performer. In this category we would include the levitation trick (who has not seen a subject left suspended prone in mid-air?), the enveloping smoke trick or even the mental telepathy trick.

But these strangers can show us one miracle of visual effect that no mortal on earth has ever achieved beyond dispute, if we discount a spellbinding influence of any sort. I refer to their extraordinary vanishing trick in which creatures or objects disappear abruptly or by degrees in open air and return by the same route. It is this marvellous feat that, probably more than anything else, confuses inquiry into the visitors' identity and almost by itself has led to a theory, called the ultraterrestrial, to explain their presence.

According to this idea — though it has various versions — UFOs come from a separate but interlocking universe which is invisible to us because its "vibratory level" is different or because we have not entered the dimension beyond ours where it exists. Presumably objects and action within that universe become visible to us only when our visitors so wish.

However, since we know so little about our own universe, it is difficult to accept this thought with all its enormous implications simply as a way of explaining one item of their magic. Maybe in time and with more insight into the paranormal it will look like a better answer, but for now the extraterrestrial theory, developed from a background of the kind pictured in the Preface by the Earl of Clancarty, an outstanding observer of the UFO scene, seems closer to the truth for most of us.

In short, we think our visitors come from planets revolving about other suns in our universe. Given the light-years of space that separate us from even the nearest of those distant worlds, it might be that over the measureless passage of time these sky travellers have created flight-ways between the stars, with satellite stopping points at intervals that make the total journey feasible. Perhaps several such journeys led to colonies in our solar system as pictured in chapter one.

There are of course, variations on this central theme. For instance, without supporting the concept of "flying saucers" one observer of the extraterrestrial scene gained attention a few years ago when he named a specific point in our visible heavens where, without any mention of established flight paths, an entire civilization was allegedly preparing to make a dash to a safer world. In 1973 Duncan Lunan, a graduate student of philosophy from the University of Glasgow, told a rapt audience of the British Interplanetary Society about research he had done on strangely delayed echoes of radio signals sent from earth into space.

On a hunch that the echoes were deliberately spaced by some means, Lunan worked out a code based on the periods of delay and was able to spell out a startling message. It said the signals were being returned by a "probe" in the orbit of earth's moon which came from a planet of a star 103 light years away in the constellation of Bootes. Lunan concluded the probe was sent by the planet's inhabitants about 13,000 years ago when, threatened by the engulfing heat of one of two suns in their system, they were forced into a desperate search for a new home. Being sceptical of the idea that ufonauts frequent our skies, however, he did not believe the probe contained a living intelligence. 'So far I've only found four 'clumps' of possible contact evidence in the whole history of mankind, and nothing more recent than 600 A.D.," he wrote me when I expressed interest in his work.

Thus we are left with a picture of a remote culture preparing for a tremendous leap into space (apparently it had fled earlier from a planet closer to the sun) as it awaited favorable reactions from the mechanical brain of its probe. Since no reply was ever sent from our end, we may never know the outcome of that ordeal.

However, if a civilization did launch itself into space for that or some other reason, its flight might have resembled what has been described as the "great meteor procession" of Feb. 9, 1913. Travelling at a height of about 35 miles, the amazing parade of lights was first seen against clear night skies over the B.C.-Alberta border in Canada. It travelled southeast, moving across the prairies and northern States and going out over the Atlantic, past Bermuda, until it disappeared off the eastern tip of Brazil, 6,000 miles away.

A Canadian astronomer who described the incident at the time was Dr. C.A. Chant. His report in *The Journal of the Royal Astronomical Society of Canada* (vol. VII, no. 3) is discussed by Arthur Bray of Ottawa in his book *The UFO Connection*. Bray says in part:

> He (Chant) states that to most observers, the outstanding feature of the phenomenon was the slow majestic motion of the bodies, and almost equally remarkable was the perfect formation which they retained. Many compared them to a fleet of airships, with lights on either side and forward and aft. Others likened them to great battleships. Still others thought they resembled a brilliantly lighted passenger train travelling in sections, and seen from a distance of several miles.

A distinct rumbling sound like distant thunder was heard in many places just as the bodies were vanishing, or shortly afterwards. In some cases, three such sounds were heard at short intervals. A number of people felt a shaking of the earth or of the house.

The number of bodies seen has never been agreed upon. Most estimates are from fifteen to twenty, but some vary from sixty to one hundred. Some people said there were thousands.

Being a good ufologist, Bray was not content to close his chapter on Dr. Chant's note that this was a meteor display, even though an exceptional one. Realizing that no other explanation was available in those days, he sought the opinion of a "well-known" astronomer in a high position, who replied:

> I, too, have been puzzled by the 'meteor' explanation and feel that, in the light of the full UFO phenomenon, this hypothesis must be seriously questioned.

In other studies of celestial activity that might have involved entire civilizations on the move we find it speculated that massive celestial bodies, natural or otherwise, were used for transportation. In his book *Our Mysterious Spaceship Moon*, Don Wilson, an American writer, enlarges on a Russian theory that provides a setting for our introductory theme. In short, the moon is a hollowed-out satellite which may have been propelled through space by refugees in ages past and finally guided into a careful circuit around our planet.

Although according to the Russian scenario that the occupants met some unknown fate, leaving the moon an empty shell, Wilson speculates they may still be on board. He remarks that among other signs of apparent lunar life, things resembling flying objects have been seen against the surface of the moon. From there we might proceed to the thought already expressed that the moon is a source of UFOs in our own atmosphere.

After meticulous research of ancient scriptures, Zecharia Sitchin, another writer in the U.S., adds to the picture of ET migration on a large scale with his theory that Earth was the result of a collision between two planets, Marduk and Tiamat. From Marduk there was a flying exodus of gods called Nefilim to a large remaining piece of Tiamat on which civilization then appeared after selective breeding with ape-like creatures performed by the Nefilim. In Sitchin's *The 12th Planet* there are extraordinary illustrations of known artifacts which point persuasively to the history he reconstructs.

Although fact must merge with myth in any search far back into time, there appears little question that great events other than those of nature were acted out in the skies during the age of creation. For that reason as much as anything else, we who favour the extraterrestrial theory to explain the presence of UFOs are loath to replace it with some idea that seems no more than a thought of the moment.

Still, the question remains: if the ET theory is correct, how do our

"energy cell": 11′x6′

visitors with their matrix of nuts and bolts manage their trick of vanishing and materializing in thin air, or sometimes being visible only on film? The answer may lie, at least partly, in the use of a seemingly rare piece of equipment which, despite the evidence of its awesome possibilities, has, somehow escaped serious attention in ufology by all but a few. I refer to the mystifying "bell-jar" device that seems to lie at the center of UFO affairs as an instant incubator of various life forms which it can also render invisible, along with itself.

As with several other subjects under study here, the matter of the bell-jar found its way into CUFOR's pages because there was someone out there with an eye for the unusual and the initiative to tell us something about it. Our contact in this case was with an observant contributor, Mrs. Betty Jones of Kalispell, Montana, who sent in a snapshot of a mountain stream running past a square-cut rock on which stood a bright blue light (see outside back cover). Anyone less alert than Betty might have dismissed the object as a shiny blue coffee percolator that campers had left behind, but when artist Brian James began making a sketch from the photo, he calculated by careful measurements that the misty canopy framing the "percolator" was six feet in diameter and 11 feet tall!

There were other oddities in the scene that became evident with careful study. On one side sat a cluster of what resembled creatures with animal-like heads. This extraordinary group appeared to be looking at the blue light. Floating above trees in the background, an indistinct ball-shaped object looked as if it were monitoring the strange tableau.

In her note about the photo Betty explained it was taken in September, 1973, at a point along the Clearwater River in Idaho. The photographer was Orval (Buzz) Wyman who noticed the attractive setting while he and his wife Irene, were returning to their home in Montana after a vacation trip. Irene had just finished taking pictures of rock formations on the shore.

"When they had the colour film developed, the rock pictures proved a bit dark," Betty wrote. "But the river scene was quite good, and it was then they saw that Buzz had photographed some unusual objects — objects they say they did not see when the picture was taken. The more they examined the picture, the more strange things they found.

"I did not know the Wymans at the time but learned of what happened when a mutual acquaintance, Mrs. Leona Nielson, sent me the river scene snapshot to obtain my opinion. I requested more information, with the result Irene came to see me, bringing two enlargements of the photo. We examined them together and found the following:

"On the river bank at the left side of the photo there is a bright blue object which I have called an 'energy cell.' It is shaped like a rounded arch and over the top is a much larger, almost transparent arch which has a bluish tinge. It occurs to me the large arch might outline the energy field of the object inside. (Irene said there were no power or telephone lines in the area.)

"The 'energy cell' itself is resting on a square rock which appears to have been manufactured or cut that way, and inside the object there are bright orange-tinted dots. The 'energy cell' is the most obvious strange thing in the picture but on closer look we could see other unusual objects on the right-hand side, facing the blue object. They look like two, three or more small grayish-brown beings, more monkey-like than human-looking, sitting on the rocks along the shore. At first glance they look like rocks but though Irene was interested in the rock formations of that area and took pictures of them, she does not remember any rocks resembling those figures."

Later Betty herself visited the site with the Wymans and took pictures of the shore-line where the animal shapes had appeared. Nothing resembling them was visible then, nor did anything similar show up in the photos.

Even at that stage, however, each of the few of us who were curious about the photo must have suspected that soon some "expert" would come forward with a mundane explanation for everything. After all, how could that great luminous object have been there without the Wymans seeing it? And what about the animal forms? Surely it was all a case of camera problems and over-eager imaginations!

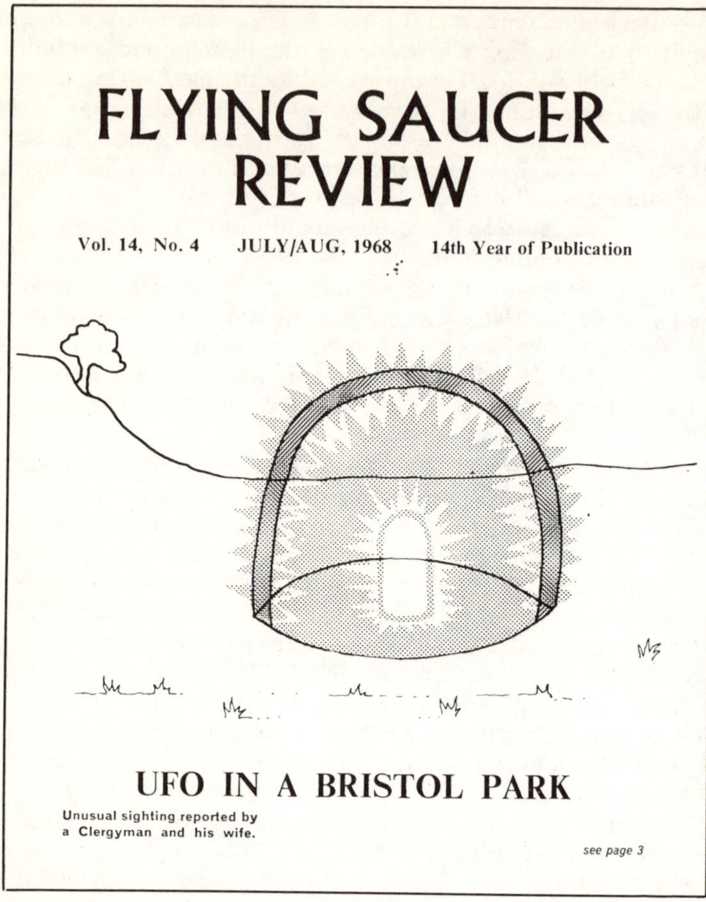

But when CUFOR carried the story, with photo, no challenge was made. Instead, the mystery deepened, for we discovered this was not the first incident involving a bell-jar, as we called the filmy object enclosing the "energy cell", because that is what it so obviously resembled.

Thanks to the good memory of two of our readers, Graham Conway and Mark Herbstritt, we learned that a mystifying bell-jar device of similar size was seen five years earlier by the minister of a church in England who reported his experience to the *Flying Saucer Review* (vol. 14, no. 4, 1968). With further thanks to editor Charles Bowen, we received permission to carry appropriate parts of the story plus reproduction of the front cover of that issue which showed an object closely similar to that seen in Idaho. The bell-jars matched each other even to the point that the large external canopy enclosed a smaller device — Betty Jones' "energy cell" — of the same size and shape. In the English case, however, the bell-jar was described as a "dirty yellow" and was seen to move, which of course added interest to the sighting.

Written by Squadron-Leader Alistair Prevost, the FSR report said:

> At 9:20 p.m. on April 27, 1968, the Reverend Anthony G. Millican, vicar of St. Christopher's, Brislington, was taking a twilight walk with his wife in deserted Arnos Park. The park is on the Brislington side of Bristol, where the suburbs of the West Country city overlap into Gloucestershire into the County of Somerset.
>
> It was a fairly warm evening after a day of continuous rain. Visibility was good, with no wind, and a clear sky with only three-eighths light cloud. There were no aircraft in the vicinity.
>
> Suddenly, without warning, the couple saw a 'glowing' object which apparently hovered some 6 feet off the ground at a distance from them of 75 to 100 yards. It seemed to be revolving, and to be maintaining a steady position, which it did throughout the sighting.
>
> This single dome-shaped object was estimated to be 12 to 15 feet in circumference. It was completely symmetrical, it neither increased nor decreased its size, and its overall shape remained constant. The colour of its outer rim was a dirty yellow, but inside the rim it was off-white and this inner light was subject to irregular variation of size. The whole thing was transparent, for the skyline at a crest of the slope beyond its position could be seen through the object. No sound was heard throughout the sighting.
>
> Unfortunately there was no opportunity for me to visit the Rev. Millican and his wife, but I am grateful to him for readily answering questions which I sent to him . . . In his reply he mentioned one curious feature, namely a central pillar 4 to 5 feet in height, but he did not state if this was solid-looking or merely an area of different light intensity.
>
> After about 20 seconds or so, the object disappeared by fading into obscurity.

Among the questions asked was whether the witnesses suffered any physical effects during or after the sighting, to which the vicar replied, "A chilling sensation. My wife felt chilled for three hours and vomited. My wife also slept very badly that night and had dreams about it. I was all right."

While the remarks were barely more than a postscript, they raised a point of considerable interest when compared to the Clearwater River case, for there the bell-jar also seemed to cause uncomfortable after-effects. Although the Wymans had no discomfort at the time of taking their snapshot, both had severe stomach cramps after their second visit to the site, as did Betty Jones, who was with them. There was nothing to link the bell-jar with their illness, but it was a coincidence worth noting that all five involved in the two bell-jar cases were affected later in the same way.

Another reader we heard from soon after Betty Jones' report appeared was our Romanian correspondent, Florin Gheorghita, who lived in the town of Cluj-Napoca. From his fund of local mysteries, Gheorghita produced one strikingly appropriate to our study of the bell-jar:

> This case happened one night in April, 1974. A young professor, Adrian Patrut, who studied the incident very analytically, sent me the following written testimony of the witness, as well as his detailed findings. The principal witness was an assistant university instructor in physics, Iuliu Marian, aged 34.
>
> "I awoke suddenly and felt immediately alert. It was after 2 a.m. My wife, who had awakened a few seconds before, was looking at the window. I looked there too and saw a strange luminous silhouette in an upright position. At the same time I had a sensation like an infusion of energy, which I cannot describe. I got up and looked out the window. The strange night visitor looked like a luminous bell-jar about two meters high. Inside was a column of light brighter than the rest of the form. After about 30 seconds I went outside to see what it was all about, though my wife begged me not to. She stayed in the room, very upset.
>
> "Outside I saw the 'bell-jar' slide quietly toward the corner of the house, remaining steadily bright. In moving to the other side it did not change direction but stayed on a course at right angles to the house. An intriguing detail was that the glowing object cast no shadow, although I could see my own shadow made by the glare of a street light.
>
> "Running to the corner of the house, I could see nothing on the other side but oddly I still felt the presence of the energy field which had probably awakened me. For about three minutes I looked around in the trees of the garden for the 'bell-jar' which I had just seen so clearly. Then suddenly I felt an overpowering urge to return to the house, and I believe there was some influence forcing me to do so.
>
> "Back in the bedroom I found my wife in a deep sleep, although she had been so nervous when I left. The next day she told me that the moment I had gone out she was overcome by a feeling of peacefulness and quickly fell asleep."

In the file of material from Gheorghita already on hand, there was reference to another incident which, in light of the Millicans' encounter, had a direct bearing on the study. After explaining how a friend, Al Sift, over a period of years had photographed glowing aerial balls in that area, he wrote:

> In the first case he remembers a similar ball landed in a tree, disappearing from view. Going to the tree and poking among the branches with a stick, young Sift had a fantastic experience — his right hand and stick became invisible for a moment! Then he felt a cold pressure coming out from that part of the tree. I shall never forget that incident (summer of 1956) which greatly changed his life.

Read in that context, and in that of other "chill" cases in UFO records, the remarks of the English clergyman about his wife's experience of feeling cold after their bell-jar sighting takes on emphasis.

Before leaving the English case, we should note this final paragraph in S/L Prevost's account referring to another incident in Bristol eight days later:

> "A woman, whose name and address were not disclosed, was wide awake and sitting up in bed when she saw a roughly circular object with a tail, like a flying tadpole, pass her window slowly and silently towards Arnos Park. It was a bright lemon colour, and apparently not so clearly defined as the previous object (of the Millicans' sighting). It was seen flying low over the rooftops for 5 to 10 seconds . . ."

Because both sightings were in the same area, with the same park being specifically mentioned, and because both incidents involved strange craft similar in colour and rounded contour, it is reasonable to suppose that the Millicans and the unidentified woman saw the same object. If they did, then we have established from the last case that the bell-jar can fly. Thus it becomes a qualified member of the flying saucer fraternity.

Now, the watchful reader will notice that having introduced our visitors' bell-jar device as an apparent incubator of various life forms, I drifted away from that point after coming close with a reference to animal shapes seen in the Clearwater River photo. So let's return to it now, and we find the place of meeting is not the shore of a stream in Idaho or the lawn of a house in Romania. It is in a parking lot in Cincinnati, Ohio.

The weird incident that occurred there in the early morning of Oct. 21, 1973 (a month after the Clearwater River photo was taken), was reported to the UFO monthly *Skylook*, now merged with the MUFON *Journal*, by Len Stringfield. It was carried by us with thanks for permission by the author and by *Skylook* editor Dwight Connelly.

The witness in the case was Mrs. R.H. (her name appeared later but initials will suffice here in deference to her original wish for anonymity), then aged 48, divorced, mother of three boys, the youngest of whom, age 13, was with her in their trailer next to the parking area in Western Cincinnati at the time. Mrs. R.H., an occasional babysitter, was described as neat in appearance and her mobile home was tidy.

Stringfield's report continued:

> "Mrs. R.H. relates that she awoke sometime around 2:30 a.m. because of intense thirst, and on rising from her bed, was attracted to a bright light coming through the drawn curtains. She pulled the curtains apart and was startled, initially, by a row of individual lights forming an arc about two yards from her window.
>
> "Each light source was as large as a hand with the fingers spread out. The lights, maybe six she recalled, were about 4 feet off

ape-like being inside "shield of light"

the ground and alternately vividly blue and silver like stars as beautiful as Christmas lights, each lit up inside (self luminous), casting no radiance to the ground or onto a nearby storage shed. The lights were in a fixed position, and did not pulsate. She did not open the window. No sound was audible, nor odor.

"While the Christmas lights still attracted her attention, the witness said her eyes were drawn next to a strong bright light further out in the asphalt parking court beyond her trailer. A car was parked on the black-top about 10 feet from the trailer, which partly obscured the bottom portion of the light source in which an ape-like creature appeared.

"Terrified, she watched this area of light for about two or three minutes and thought the creature appeared to the rear of the parked car maybe doing something to the car. (It was at this time she ran to her son's bedroom and tried to awaken him. She yelled, My God, Carl, get up . . . get up and look at that thing out there!)

"When she returned to the window, Mrs. R.H. relates, the creature appeared to be further away from the car, maybe 35 feet away. This time it was inside a shield of light, the light looking like a light in an operating room. The shield was shaped like a lady's umbrella, the kind that comes down over the head (often referred to as a bubble umbrella). The bright light was within the shield. There was no glow of light beyond the shield.

The creature "stood out clearly," looking like "an ape" with a big waist and no neck. There were no distinguishable features to the head, but Mrs. R.H.'s description indicated it was a side view of a head of prognathic configuration with a snout, which tapered downward. The head and body were facing the Swallen's warehouse building, which was to the left of the witness' mobile home. The body was entirely gray and was also featureless. Only the arms on each side of the big waist were moving "very slowly" alternately up and down.

Mrs. R.H. showed me the up and down motion without bending elbows, which appeared to be the characteristic stiffness of a robot. No other part of the body moved. At one point, Mrs. R.H., in trying to describe the object's shape, held up a bell-jar. "It was like glass," she said, "a glass shield that had a bright light inside it. It was round on top."

The object's circumference was estimated to be equal to "six men each holding hands with their arms outstretched, standing in a circle." She added, "there was plenty of room inside the lighted dome to hold several more creatures of similar size." By Mrs. R.H.'s estimate, the object or shield was approximately 7 feet in diameter. There were no visible instruments or panels inside the self-luminous capsule, although the occupant's arms appeared to be moving a lever that was not visible. Mrs. R.H. related that the humanoid remained in view about four or five minutes during the whole episode.

Stringfield went on to report that according to Carl's testimony, he saw the light his mother indicated while still half asleep. He also saw a car arrive with two young occupants who tried to start the parked car. Later it was learned the parked car had chronic engine trouble, so no connection between that problem and the rest of the incident could be established. However, it was also learned that Swallen's warehouse fire alarm was tripped just at that time by an unknown cause. Meanwhile Mrs. R.H. was vainly phoning the police and, at some point during that period, the light shield and its occupant disappeared.

In this account we again have details of stunning similarity to those in the other cases. When Brian James sketched the light shield with an animal inside by carefully following the information given by Mrs. R.H., he established that this was indeed another case involving a huge bell-jar shape. The rounded top, like that of a bubble umbrella, gave it the jar's appearance as the witness herself described, and the diameter of seven feet was within a foot of that estimated for the Clearwater River object in Idaho. (In the English case I suspect the minister may have underestimated when he put the circumference at 10 to 12 feet. That would have meant an object with a diameter of about four feet, much narrower than in the sketch based on the witnesses' impression.) Judged in relation to the parked car, the shield must have been several feet taller than a normal man as in the Idaho and English cases, and the snouty appearance of the occupant seemed to put it in the same category as the "animals" by the Clearwater River.

However, just as important as similarities in this case is the fact that the creature was seen *inside* the bell-jar. Further, since it was in surroundings utterly alien to a beast of that appearance, it is not too much to infer that the creature depended on the device for its very existence, even though it was seen outside for a moment. While it was inside the shield of light, the motion of its arms might well have been connected with the operation of something to sustain its life, and its obvious reliance on that light casts our thoughts back to the Clearwater River incident. Were the creatures there watching the blue light intently for the arrival of another of their kind?

If the idea that the bell-jar device is an incubator for strange instantaneous forms of life seems too extreme, let's consider the experience in 1974 of William Bosak, then 70, a respected dairy farmer of Frederic, Wisconsin, who encountered an unusual creature on the highway near his home. This incident involved a large container with a "curved front of glass" which Bosak called a "vehicle." Reading about it in a clipping from the St. Paul (Minn.) *Pioneer Press*, I saw at once that the "vehicle" drawn for the paper according to the witness' description, resembled that now familiar device, the bell-jar. Intrigued, I wrote Bosak for details and received this useful reply:

> I was coming home from a co-op meeting at about 10:30 p.m. It was the early part of December and very mild for the time of year. There were patches of fog on the road so I was driving slowly with my car headlights on low beam.
>
> When I was about one half-mile from home I noticed something in the left-hand lane of the road, so I slowed up nearly to a stop. When I was only a few feet from this vehicle I could see it very plainly. I could only see the top part of the vehicle and the occupant from the waist up.
>
> The strange being in the vehicle had his hands up as if to show he was surrendering or to show that he meant no harm. His eyes showed intense fright.
>
> He was a very strange looking man. His hair on the sides of his head stuck straight out about an inch, but he did not have hair on his face like a beard or whiskers. I tried to see what clothes he was wearing but I believe the fur that covered his body was really his skin and not a fur suit because I could not see a seam down the front or a collar around his neck like there is on a shirt or jacket. The fur was a sort of reddish brown or a bit darker than what we know as Guernsey colour. His ears were a more like a calf's ears than like human ears and they stuck straight out at least three inches.
>
> This person was slender and I think that he was about six feet tall but he seemed much taller because I suppose the floor of the vehicle was about two or more feet up off the road. I could not see

the lower part of the vehicle because it was in the fog.

He seemed to have quite a flat look to his face, so with the strange ears and hair sticking straight out it made it pretty frightening. Besides he seemed so tall. I also noticed his arms as he stood there with his hands up, and there was hair sticking out on his arms. I could not see any end of a sleeve so I do believe the fur was really his own skin. I did not see his hands very well but he held them right together as he held them up.

When I got right alongside the vehicle which was about six or eight feet away, he was watching me. As I passed, it seemed the object came right toward my car, and it became very dark in the car. I was looking back as I passed him and I do not know if the car lights dimmed or what caused it to become so dark in the car.

When this object did take off as I passed, there was a sort of swish and it seemed as though something brushed against the car. It did definitely seem as though it came right at me and there seemed to be a tremendous surge of power.

I did not tell you at the start that the first thing I thought of when I saw the object was, 'now what in the hell is that'? I also knew that if I ever told of this, or even dared to tell, I'd have to describe it very thoroughly.

At the end of his account Bosak answered certain specific questions I had asked him. One of these was to determine what the occupant mostly resembled, man or animal? His answer suggested man.

> I could not say what colour his eyes were but they were eyes like that of any man, he wrote. "I could not remember the exact shape of his mouth and nose but his neck was moderate in length and like a man's neck. His head was about the size of an average man's head.

Asked for more details about the object enclosing the creature and how he fitted into it, Bosak replied:

> The transparent enclosure was about six feet across, so the occupant did not fill it. He was about six feet tall and slender. The container was standing upright (or hovering?) in the left lane on a blacktop road. There was no snow on the road. I could only see the object above the fog, which was about at the occupant's waistline. He did not move toward me but stood there with his hands up and looking possibly as frightened as I was.

(Note the enclosure's diameter of "about six feet" making it the same as that of the bell-jar pictured in Idaho.)

Although the creature itself evidently did not move, it appears from Bosak's earlier remark that, as he passed, the enclosure rose with its occupant and almost brushed against the car.

The witness explained that despite a low-lying fog he could see what was happening because the lights of his car illuminated the object. "It did not have any lights of its own that I could see," he added.

Before mentioning his experience to a local reporter, whose account was then picked up by the St. Paul paper, Bosak debated with himself for several weeks about what he should do. His hesitation was understandable. Operator of a 450-acre farm for 40 years, he was staking against his story his reputation as a responsible, public-minded member of the Frederic community. The risk was considerable but, as in many other cases, the impact made by his remarkable sighting was difficult to ignore. Perhaps the real question facing a witness in that predicament is not so much whether to endanger his good name, however distasteful that may be, but whether to be honest with himself by saying something incredible he knows to be true.

For Bosak the problem might have been easier if he had known there were to be other strange encounters through 1975 in that same area of northwest Wisconsin. As events unfolded it seemed as if our UFO visitors had some special project underway, to judge from the number and variety of sightings. Possibly the creature seen in the bell-jar on the highway was just part of a larger plan or possibly he was the reason for all the fuss. After all, placing a new intelligence on earth, as the bewildered part-man appeared to be, must be a major effort even for the gifted aliens.

Now we will have a closer look at that activity and follow it with a still closer look at the bell-jar.

Chapter Three: The Big Wheel

Brian James

A hint that something was afoot as a sequel to the Bosak incident came when a report appeared in the APRO *Bulletin* (June, 1975) that Mrs. Ruth Christiansen, also of Frederic, Wisconsin, had seen unusual tracks in the snow outside her home. Containing imprints like small horse-shoes, the tracks could not be identified by Mrs. Christiansen, even though she studied wildlife and was enough of an expert to write magazine articles about it.

There were two points of immediate interest about that report. First, the place of origin, Frederic; second, those imprints resembling small horse-shoes. The interest of the first point was obvious as here we had Frederic once again the center of a mystery. But what about the second? What bearing did tracks of that particular shape have on the bell-jar affair?

To understand there could be a connection, let's go back over the years to a February night in southern England in 1855. Although nothing unusual was heard, something crossed the snow-clad countryside that night that left tracks running in a straight line for 40 miles. Because the tracks looked like hoof-marks, each about four inches long, there was a rash of letters in the press speculating on what manner of animal was involved but none could say why the single-file trail was so straight and how it climbed over everything in its way. Eventually the Beast of Devonshire quietly took its place in the records of the unexplained. That, however, did not mean it was forgotten. When ufologists came along about a century later there was more ado about the Beast because certain tracks in a variety of sightings suggested the Beast was still on the prowl, and it was clear now it had some sort of association with UFOs.

By chance, Bill Allan, a dedicated student of the UFO mystery, and I were in the vicinity of one of those sightings soon after it occurred and we were able to talk to the principal witness while his impressions were fresh. It happened one afternoon in September, 1971, in the fossil and oil grounds of Drumheller in the "badlands" of Alberta. Apparently, as part of their program to make underground surveys, our visitors favour the Drumheller area, and this time what was supposedly one of their vehicles had made a landing.

We spoke to the oil company's supervisor, Joe Klimek, about it. He said that with two others he saw a squarish metallic object measuring about 12 by 12 by 10 feet just below the lip of a canyon where no structure should have been. Unequipped just then to climb up or down to the ledge where the object stood, Klimek drove back with his companions next day for a closer look. The object was gone.

"There is no way a thing of that size could have been lifted out of there without making it a major operation," Klimek said. "Even if a helicopter had managed to do it, there would have been tracks all over the place, getting the thing hitched up. But there was nothing like that at all. If it didn't fly off under its own power, I don't see how it got away."

As it turned out, there *were* tracks but such unusual ones they only increased the mystery. Klimek discovered them when he climbed down into the canyon. Studying the clay slope just above the ledge, he saw a series of oval-shaped marks about six inches long leading in a *straight single-file line* to a grassy bank where they disappeared.

By themselves those tracks suggested virtually nothing about who or what was responsible, but they did not stand alone. About that same time I read of an earlier incident at Custer, Washington, when a circular mark was found in a snow-covered field where a UFO was believed to have landed. Leading away from the circle in a dead straight line were oval-shaped prints that went in single file to a clump of evergreens where they disappeared. The prints were eight inches apart, exactly the same as those left by the Beast of Devonshire.

By now straight single-file lines and equally spaced oval-shaped prints were beginning to add up to the track of a mechanical contraption of some sort. So, to illustrate an article we did on the Beast, I asked artist Brian James to sketch a single wheel with pads on its rim moving under its own power across a snowy expanse of Devonshire and leaving behind a single file of "hoof" marks. While imagining it as a

UFO device that might explain the identity of the Beast, I did not dream such a thing would ever be seen.

As it turned out, however, I was late by several years. My awakening came via the *Flying Saucer Review* in which I had read something about remote-controlled objects. Hoping there was more, I wrote editor Charles Bowen for help, and hit the jackpot. In a back copy he kindly sent me (vol. 12, no. 5, 1966) there was an article that Bowen himself had written titled *Cross-Country Cog Wheels*. In it was an excerpt from the *France Soir* of March 31, 1966, which said:

> Here from another world, a sonde device designed to record the details of the relief of the Earth was possibly strolling around in the Vosges last week-end. This extraordinary hypothesis is offered by the investigators who led the job of explaining the presence of a mysterious 'wheel' near the village of Tranqueville-Graux.
>
> The machine was seen by a shepherd, M. Jean Voilquin, aged 54, a serious, well-balanced man incapable of perpetrating a hoax, so say not only the gendarmes but also his employer, M. Mangeot.
>
> "On Saturday at 13.30 hours," said the shepherd, "I was watching the sheep at a place about two kilometres from the farm. Suddenly I saw a sort of vertical wheel coming towards me. It approached rapidly at a speed of 25-30 km.p.h. When it passed level where I was, it was 15 metres from me. It measured about 80 cm. in thickness. Its circumference was red and had cogs. It followed the terrain perfectly: each time a cog touched the ground there was a slight sound. My two dogs set off after it, but its speed was absolutely constant and did not vary when it crossed a slight dip in the ground. It was about 600 metres from me when it disappeared."
>
> Last night the enquiries had still not been explored. The hypothesis of the "sonde from another world" would perhaps explain the external cogged effect similar to a caterpillar tread and registering on a seismographic sheet all the changes in level encountered, however tiny.

So there it was! The Beast was literally a Big Wheel on the UFO scene, perhaps as important in its way as the bell-jar. With this in mind, let's turn to Mrs. Christiansen in Frederic, Wisconsin, who saw the tell-tale tracks in the snow outside her home. There the "hoof" marks were smaller than in the Devonshire and Drumheller cases (no tracks were found in the French incident, and no measurements were recorded at Custer, Washington), and followed an erratic course in places. Yet their similarity in shape, plus the fact that straight single-file lines and eight-inch spacings did show up in the snow patterns, strongly indicated that at least a junior version of the Beast had been in the Frederic neighbourhood. Just two months earlier the bell-jar with its captive humanoid had been there as well.

What was it all about? Hoping for another clue or two, I wrote Mrs.

Christiansen and found instead that the plot thickened. While the photos she sent in reply showing prints in the snow did confirm they were unlike any with a normal explanation, it was what she added that spelled out more of the riddle centered on the rural community of Frederic that winter.

After explaining that the time in question was early February, which was also when the tracks had started to appear around her place, she wrote:

> My husband had gone to bed but I was sitting up later, my usual habit since I do most of my freelance writing at night.
>
> It happens that a little bookcase stands under the north living room window, which is small and high on the wall. At about 11:00 p.m. I chanced to glance out this window. My attention was drawn to a strange double light, definitely not a car light.
>
> The lights were a deep red, about the size and shape of dinner plates, one sitting on top of the other. My husband had gone to bed, but I turned off the house light and called for him to come and look. The lights did not move or flicker, or give out a glow, and seemed to be in the road up the hill aways. We stood by the windows for perhaps 10 minutes or more, he at the little high window and me at the double windows of the dining room.
>
> Suddenly a strange, frightening thing happened, right before my eyes. Directly in front of me, outside the house, about four feet from the wall, there flashed a terrific light, blinding like a camera flash light but many times brighter, and something like a fire ball, about four feet across or bigger. It was either on the ground or near the ground, possibly a foot above the ground.
>
> I shouted, "WHAT'S THAT?" The light did not go out but instantly reduced itself in size to about two or two and a half feet in diameter. I could see clearly an object of this description:
>
> An intense gob of light, nearly like fire, engulfed a basket-ball- sized object. From the front end emitted a rapidly pulsating light, upward and slightly to the back. The light was blue-white with pinkish fringe at the back. The engulfed object was located at the rear, and appeared round, dark in color, such as black or deep blue or green, but not red.
>
> From a standstill, the fiery light suddenly sped at a terrific rate, along the wall and towards the west. The yard is full of trees and shrubbery and a snowmobile could not have travelled there without hitting something in the dark, or even in the daylight. But there was no sound whatsoever, and the fiery scooter was not long enough for a snowmobile. Anyway, I would have instantly recognized such a vehicle.
>
> My husband was standing too far from the tall window, and looking up towards the lights (double light) on the hill, and did not see the fiery light or object. The double lights remained on and

were still on when my husband went back to bed in about 10 or 15 minutes.

I sat up until 2:00 p.m. and during this time turned the house lights off to watch many times, up the road. During this time several strange things happened.

About every 15 or 20 minutes I turned off the house light to watch out the little high window. Sometimes when I went there, the double light would suddenly appear and remain on all the 15 minutes I watched. Sometimes it would already be there, but would go out, then come back on after several seconds.

At one time a different light appeared, seemingly about 10 or 12 feet from the double light (while the double light was on), and looked about washtub size. It was deep red like the double lights. It remained on for probably 10 seconds and I saw it only one time. There was no glow to any of the lights, just like red discs of light, no motion or wavering, and always in the same position. Sometimes a car sped past, and then the double light turned off. The cars did not slow down or turn out, as if for a parked car along the road.

Then, probably about 1:30 a.m., a startling thing happened. I had turned off my house lights and was watching up the hill into the darkness. Suddenly, a deep red floodlight appeared, under a vehicle! It appeared in the same place as the double lights, which were not there just then. The flood light was bright enough to show snow under the vehicle, and I then realized the vehicle was positioned along the side of the road and not in it or on the shoulder.

"The vehicle was not on the ground, but raised above it, seemingly about two or two and a half feet above the snow. It was squarish, very much like a wagon box, with the wheels or landing gear on the outside of the body, not under like a car. I could clearly see three wheels.

I could see about two feet of the bottom of the vehicle, and above that was all darkness, so I have no idea as to how tall it was. It was smaller than an ordinary car, and definitely was not an automobile. It didn't look like a helicopter, either.

I thought about putting on my coat and boots and investigating, but when the floodlight went out in about half a minute, and all was darkness, I gave up the thought.

The double light appeared once or twice after the flood light disappeared. I saw it there at 2:30 a.m. for about 15 minutes (still there when I left the window), but I did not see it at 3:00 a.m. when I went to bed.

About two weeks after we sighted the double lights, our neighbour who lives one-quarter mile to the south of us, said he had also seen the double lights, exactly as we described them. But he is a 'non-believer' and said they had to be a reflection. But a reflection would not have come and gone as we saw the lights do.

Fortunately for us at a time of such unusual activity in her part of Wisconsin, Mrs. Christiansen became a good friend and, with an eye for what was newsworthy, kept us well informed with notes and clippings about what was going on. As 1975 ended, the local *Inter-County Leader* brought mysterious matters up to date with this report, quoted in part:

> More recent oddities include a report from Gunnard Linder of Frederic. He reports seeing a large, slow blinking red light, about two feet in diameter just over the tree-tops in the direction of the Callaway residence. It was red in colour and appeared about 5:30 p.m. several weeks ago, just before Christmas.
>
> His dogs saw it too, barking excitedly and drawing his attention. As he stood and watched for some time, there was no sound, but the light appeared to be moving very slowly and finally dipped down and was out of sight. He has not heard any explanation.
>
> Sightings of strange, hairy creatures have also been heard from others. Olander Jensen of Clam Falls was driving home from a meeting in Frederic one night in December when he saw "two big fellows with long hair and beards" alongside the road near the Louis Johnson Place.
>
> As his car approached, one of the two walked out onto the road and waved his arms as if to signal the car to stop. But Jensen did not stop. Although his impression was that the men were probably just long-haired bearded humans, the Bosak report of the same location gave him some second thoughts.
>
> The most recent report is from another resident of the area east of 35. Jerry Strese, who is employed at Dresser, reports that he was on his way to work, before daylight, one morning after Christmas, when he saw a hairy creature run across the road in front of him.
>
> He was travelling the town road which leads south from "Sullivan's Corner." Suddenly a huge body, about eight feet tall, came out of a swampy area close to highway 48. With long hair, and appearing to be furry all over, the creature ran, on two feet, in a hunched-over position, real fast, across the road.
>
> The fur appeared to be rather light in colour, but the body, otherwise seemed manlike. Strese continued on his way to work, somewhat "shook" by the experience, and is still without an explanation for the strange sight in the early morning hour.

From the information supplied by Ruth Christiansen it was apparent that her section of Wisconsin and that of Bill Bosak, a scenic area of lakes and fast-running rivers called Polk County, had something of a history of bizarre events. Well reported sightings of UFOs, for instance, went back several years.

Obviously, then, Polk County has something of particular interest to our visitors. Whatever it may be, the gathering of the aerial clan was there in force in 1975 and Bosak's encounter just before the ceremonies began seemed to give the remarkable bell-jar device an important part in those proceedings.

Its full purpose, of course, remains a mystery like everything else connected with UFO affairs. But the creature on the highway who appeared to be a confused captive of the instrument, coupled perhaps with the other sightings of hairy figures, again leads us to think it is able to generate various forms of instant life seemingly in a fully grown state. Also since there was other traffic on that blacktop road, yet Bosak was the only person to report his experience, we can assume that quick disappearance was achieved as well.

So far, however, in this study of the bell-jar, we have considered cases that involve few participants — at most, the number of figures seated around the instrument by the Clearwater River in Idaho. While the mystery of abrupt appearance and disappearance is evident in every instance (allowing for the fact that in the Idaho case all the action was concentrated into a single camera shot), it might be said we should offer something with a broader setting to compare with the magic considered possible under the rival ultraterrestrial theory. That theory was cited early in the preceding chapter as one way of explaining how UFOs may come and go through a veil of invisibility. In other words, the bell-jar might explain some of our visitors' performances at the ET level, but do the effects of that performance match what might be seen on entering another universe?

While that, of course, is a question no one in the world is qualified to answer, I do have on file a case involving the bell-jar form in which the witness either entered or saw what we might *suppose* was another universe or dimension. The report that follows on this remarkable case was contributed by an outstanding ufologist in South America, Irene Granchi of Rio de Janeiro. Besides her numerous other activities in UFO organizations, which included serving as editor-in-chief of the Brazilian periodical *OVNI DOCUMENTO*, Mme. Granchi was a correspondent for CUFOR for several years and thus became interested in our study of bell-jar cases. Soon she was sending in reports of her own on the subject, a high point of which was "Antonio's Ordeal" as we called this fascinating account.

While Antonio's case was of interest generally in the varied field of UFO abductions, it was particularly appropriate for our purpose because after he was captured by ufonauts everything that happened to Antonio did so while he was inside a "redoma" which, translated, is "bell-jar." This meant that here we had the extraordinary case of a human in a captive position similar to that of the animal in the Cincinnati parking lot and the humanoid on the Wisconsin highway.

What Antonio experienced in that position was so utterly strange he might well have been in another realm of time and space. Yet externally everything else, including his first sight of the aliens' craft, had taken place in reasonably normal surroundings. After his release

Antonio evidently saw the craft fly off above the town where it had landed in the first place.

Thus we may conclude that while the UFO itself was operating within definable areas of space, what Antonio observed through the medium of his "redoma" extended beyond those limits into what was possibly another dimension. Additionally, much of what he saw was projected on a screen as if yet another medium was required to complete total reception from that extra dimension, if such it was. Images of himself in situations he did not recognize further suggested that an extra dimension was indeed involved.

Even so, we cannot sensibly say our visitors must come from that nebulous domain. Rather, it may be supposed that if the bell-jar serves as some kind of receiver that can pick up scenes from way out "there", the aliens may pack it along on their trips from the stars just as they carry so much other magic-making luggage. Admittedly it is difficult to see how this video instrument, remarkable as it is, can double as an incubator and vanishing device, which obviously it does. But, then, who said anything in ufology was simple?

Admittedly, too, it may be difficult for some of us to visualize a complex instrument of this kind that could measure up to 15 feet in height, as it did possibly in the Bristol sighting of the Millicans, being packed aboard a much smaller UFO. But let's remember our visitors are nothing if not tourists (as well as magicians), and evidently they know the value of "inflatable" articles, as we gather from an incident that occurred in a town near Brussels more than 100 years ago.

A report of this appeared in the excellent Belgian UFO periodical *Inforespace* (no. 2, 1972). The witness was Leon Herbosch, aged about 30, who was out for a walk in Rhode Saint-Genese one evening in January, 1870, when he noticed a greenish "blob" of light on an empty site by the street. The item continued:

> Within a period of about 20 to 30 seconds, the blob became increasingly brighter, seemed to vibrate, then rose and "inflated" into a goldish bell-shape. This "luminous bell" enlarged and reached about five metres in height. Its luminosity became almost blinding and its colour paler, almost a "neon white".
>
> The light, although vibrating, retained the sharp bell shape, its borders being a little like a ray of light — very sharp but not limited to precise place. Its interior seemed made up of tiny luminous particles swarming all around. Once at its maximum height, the object lit up.
>
> Extremely frightened, a few seconds later the witness saw the bell suddenly 'falling down or going flat' to resume its original shape. The blob then moved away . . . exactly following the contours of the ground . . .

(Note the reference to "tiny luminous particles." In describing the "energy cell" contained in the bell-jar photographed beside the Clearwater river in Idaho, Betty Jones, as quoted earlier, said "inside the object there are bright orange-tinted dots.")

In coming now to Mme. Granchi's report, we will see that all Antonio had to show for his unusual experience was great physical distress. While it may be felt that in places this is described in too much detail, the purpose is to underline the reality of what happened to the witness. If everything he went through was as unreal as some of the scenario suggests, it seems unlikely he would have emerged in such an obviously sick condition.

Mme. Granchi prefaced her report by explaining that the witness, Antonio Bogado La Rubia, lived at Paciencia, near Rio de Janeiro, with his wife and small son. She said he had received an award from his employer for attentive service as a bus-driver.

Then in a follow-up note on the case, she wrote, "Unfortunately there is nothing to add to the story as Antonio has gone into anonymity after he found himself a job as a private chauffeur, where he did not wish anyone to discover what had happened to him. I have his in-laws' number, but they must have moved, as nobody replies on calling."

Opening with an excerpt from the newspaper *O Dia*, Oct. 4, 1977, Mme. Granchi's account of "Antonio's Ordeal" follows in the next chapter.

Chapter Four: Antonio's Ordeal

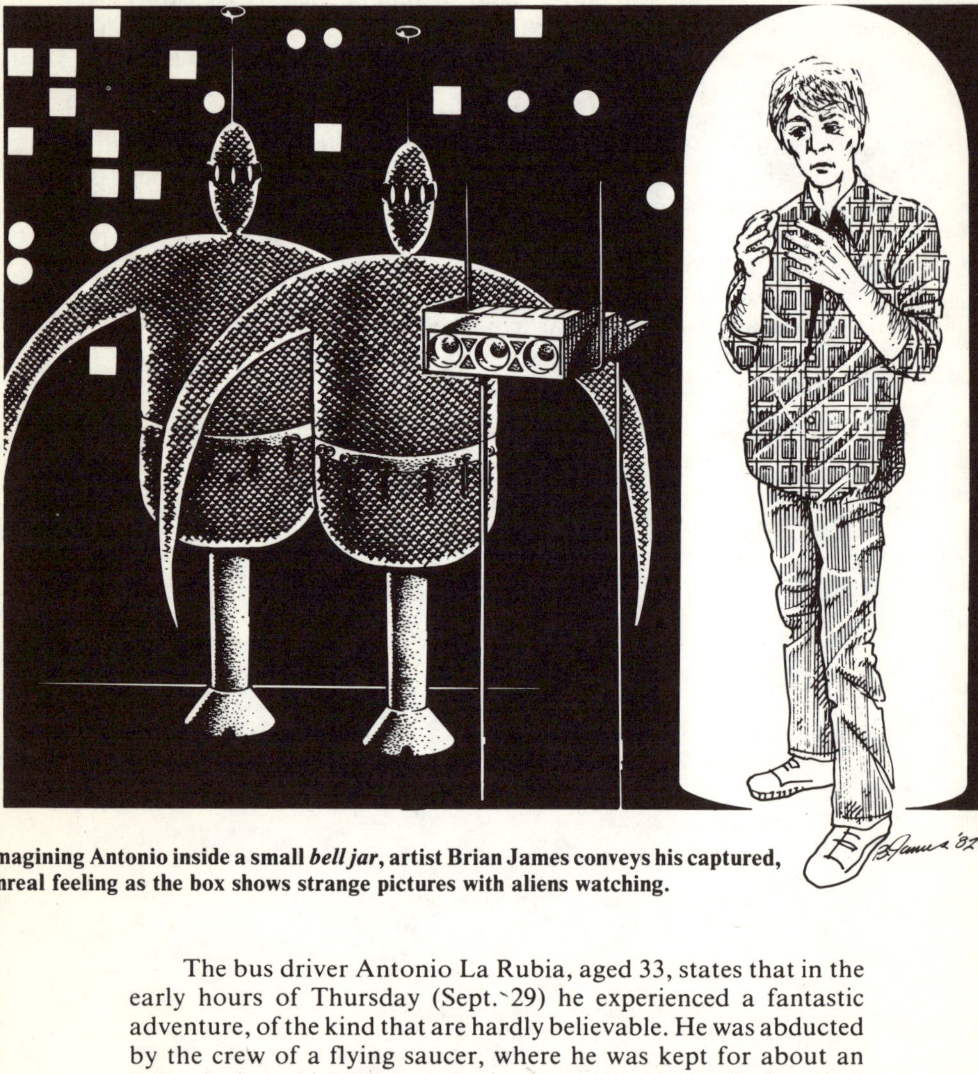

Imagining Antonio inside a small *bell jar*, artist Brian James conveys his captured, unreal feeling as the box shows strange pictures with aliens watching.

The bus driver Antonio La Rubia, aged 33, states that in the early hours of Thursday (Sept. 29) he experienced a fantastic adventure, of the kind that are hardly believable. He was abducted by the crew of a flying saucer, where he was kept for about an hour, during which he was submitted to several tests. He described all he could, and he presents really strange symptoms, for at intervals his temperature rises to 42 deg. C., as was checked by the nurse, Estelita Ferreira Pereira and the psychologist Neli Carbonell David, both employed by the Oriental bus company where the patient works.

"Inside the disc, it seemed as if they wanted to roast me alive", he said.

Antonio left home at 2:15 on his way to the highway where he would be taking the 2:35 bus. From there he would continue to his company's garage to fetch his bus for his daily round. He was feeling calm and his only fear was that he could be assaulted, as the spot is deserted at that hour.

Going by a little square, he saw something that made him recoil: a flying saucer. He says: "It looked like an enormous hat. I only took two steps back, for then I was nailed to the ground and saw two little men grabbing me. I cannot recall how I entered the disc, but suddenly there I was, floating between two rows of a dozen little men on each side. They were small, but had large shoulders. Their heads were like those American footballs. There was total silence."

Antonio panicked, and a shout was strangled in his throat; for some strange reason it did not come out. He described all he saw, including the fact that the place was profusely lighted in blue and white. "There were only those strange little men around, about 50 of them. I felt as if I were in a crystal bell-jar and felt that they were communicating among themselves, as their heads turned round when they were saying something. They kept still and watched me with fiery eyes. There was a screen, where they showed me a horse drawing a cart, a dog barking and a sudden flash that killed the animal. On the screen I was also shown a countryman in a straw hat. I do not know what all that meant, but I was all the more surprised when I watched myself on that screen. I was naked and they were studying me." He said he gathered all his strength to shout, and asked: "Who are you? What do you want from me?"

At that moment, according to him, it looked as if those men were the victims of some mortal weapon, as they fell down to the floor. Just as suddenly, Antonio's voice failed him. He was paralysed by the energy of a beam of light directed at him. "It looked as if they wanted to roast me alive. That beam of light turned me red, burning like fire, and I was overcome as it lessened."

In the same manner as he had been taken into the disc, Antonio was released from it, and he does not remember how, but he found himself in a lane near where he had been abducted. His head was spinning, he was burning, and all he can remember is that he asked a passerby the time. It was 2:55 a.m. As soon as he reached the firm, he asked his companions to turn the water-hose on him, as he was burning so. They turned the bus waterhoses on him and that helped. "I was taken to the Infirmary and the nurse, Estelita Ferreira, discovered my temperature was 42 deg. This went down, but soon after the temperature went right up again. I

Antonio (right) describes his capture to a friend.

refused all medicine, and was afraid of taking an injection, for fear of the reactions it might produce."

Later, he was taken to see the firm's psychologist, Dr. Neli Carbonel David, who gave our reporter the following statement: "I cannot say anything yet, but the patient is very nervous Clinically, his state was rather strange as his temperature was so variable. Psychologically speaking, his state is disastrous. I shall take some more tests of him tomorrow, but they will have to be submitted to a medical team, for his clinical state is fantastic."

Antonio is staying with his mother-in-law, and he let himself be interviewed in the corridor of the building, for he is afraid of his wife and mother-in-law hearing the true facts. He kept it all from them. But sooner or later he will have to speak, as the burning comes and goes. He is also afraid that what he saw on the screen inside the flying saucer might in fact come to pass with him, but he did not say what it was. All he wants is to forget everything, and be cured of those strange lesions.

The newspaper went on to say that they interviewed a doctor as to the high temperature Antonio ran, who said it was usually fatal in an adult.

There follows my own (author's) account, with the facts gathered from the bus company's psychologist, Neli Carbonell (she does not wish to be called David) and the witness himself, Antonio.

During my phone call of Oct. 6 to Dr. Neli, almost the first thing she told me was that she does not believe in UFOs, and promised to give me her reasons for this. She is the psychologist employed to select personnel at the Oriental bus company, the richest in the outer regions of Rio. She informed me that Antonio had been swollen, had suffered from loose bowels and vomiting. She did not think it could be the effect of radiation poisoning, as I immediately suggested, but had not had any tests made to that effect. She was clearly nonplussed by the whole case, and we made an appointment for Saturday morning, in Campo Grande, at her consulting office.

Pamela Talbot, who has so often helped me in the past, drove me out from Rio, and we got there shortly after nine. When Dr. Neli arrived, her major concern was to make it clear that she wished to keep Antonio's spirits up, and needed him to improve emotionally speaking, so all sensationalism was to be avoided. She was pleased to be able to tell us that Antonio had passed his first set of tests beautifully, with only one unsatisfactory reply and 20 satisfactory ones. She had applied the Rorschach test, and would proceed to other tests with a team this coming week.

Although her statement is not final, her impression is that the man is normal. He has a high I.Q., and is very duty-conscious. Dr. Neli confirmed that he had had high fever, but when Antonio came into her hands, it had already gone and did not return. When she first saw him, he was stark naked, swollen, with a 'brotoeja'-type eruption all over his body (brotoeja — rash) and scratches, which he said he had made himself, because of the itching. He would not allow anyone to touch him or to clothe him. She confirmed that he had vomited and suffered from loose bowels.

She finally persuaded him to get dressed, and then he gradually improved. His thirst was boundless, and his mouth was still burning. The hematomas appeared with little eruptions at his joints and on his neck. He has recovered from all this by now, and his skin is clean, though he is still very nervous. Much less nervous though, than he was at first for, when Dr. Neli first saw him, Antonio was crying like a baby, in a pitiful state.

She feels very sorry for him, for he has a family to keep, and a sick mother. She had already asked him, at my request, if he wanted to see me, but he refused, saying he was tired of repeating the same story and would see me some time in the future. He was taken to the Rocha Faria Hospital, after the infirmary, and there the specialists said this was *not the first case* of the kind that had come there, and that his case was very delicate. In some of the former cases, the patients had gone mad.

Then Dr. Neli gave us a short account of her patient's story, describing the physical appearance of his captors in detail, and even making an (unsigned) sketch of one of the 'dolls' or 'robots,' which I later checked with the very sketchy account Antonio made. It was slightly different, which only goes to show that even psychologists are subjectively swayed by their personal concepts. But her general outline confirmed what the newspaper had said. I could see she was much more impressed by Antonio's story than she cared to admit. She did not wish me to say I had been to see her if I went to visit Antonio, which I intended doing.

We proceeded to Paciencia, and it took some time to reach Antonio's house through mudlanes that were hole-ridden. We found it closed, and discovered from a neighbor that he was still staying with his in-laws, so we turned back. The neighbour also pointed at the field where the UFO sighting had been made, and we stopped there. It was too large for me to examine from end to end, the grass was the low creeper-style, so there was nothing wind-swept on it. There were some youngsters around who did not look too friendly, and the region is well known for crimes and assaults, so I had to be careful not to stay too long.

I took two or three pictures of the surroundings from this field. These were interesting: at the extreme end of a long plain there was hilly countryside, at the bottom of which ran the high power-lines of Furnas, which furnish Rio with electricity, and have so often appeared in my past reports. Nearer us, there were transformers and more powerlines. Otherwise a deserted landscape, with a few rows of houses and lanes crossing the plots.

We finally discovered Antonio's present dwelling, an apartment on the first floor above a bakery, and found a very pleasant, friendly family, who persuaded Antonio to come out of his room and meet me. Antonio is a goodlooking, muscular man, tanned, blue-eyed, with light chestnut hair. He has a ringing voice and looks well-balanced. His wife is pretty, and their child is beautiful. Antonio was afraid from the first that I was another reporter, and would not speak. He is now a chain-smoker, and still very thirsty all the time. He asked me not to tape-record anything, but I tried to cheat, and quick-witted as he is, he soon discovered this, so nothing was tape-recorded. (I tape-recorded the psychologists's interview though, with her permission. I do not usually cheat.)

Antonio said the newspaper had written the truth, but that the beings were not small, they were almost his height, and he must be about 1.66 m. (5'5½"). They wore metal helmets, rugby-shaped, with a band running across it where it was broadest, all cut into mirror-like sections, from which blue flashes were emitted that grew fainter and brighter and had a hypnotic effect on him. (I say the word hypnotic, not he.) They were metal-clad, but though they seemed to be robots, they breathed; he could hear their breathing.

When I tried getting confirmation of the description made by the psychologist he stopped short, saying again and again that I should read the newspaper, that some day in the future, after he felt better, he would give me the full story, but he did not feel like it now. He confirmed that there had been over a dozen of the beings on each side of him inside the UFO, where all was blue and white, and that blood had been drawn from the middle finger of his right hand, of which Dr. Neli had already told me. He chose UFO 4-C in the USA 1968 Hearings of the Committee publication, saying it was the nearest thing but not exactly what he had seen.

Every time I tried to obtain more details, I failed. He also confirmed there was absolute silence, and that communication was made through the TV-like screen, where they showed him a dog being killed, then one of themselves being killed (according to Pamela, and I agree, it was an object lesson showing that they kill ours as we kill them, or something to that effect). The screen also showed him naked surrounded by doctors, exactly as really occurred with him later, and also again of himself wearing clothes and dirtying them with his stools, but this did not occur, as he was able to retain himself when he in fact became so ill. Then another scene, which he refuses to describe.

Antonio said that while he was inside the ship he felt well, normal. It was only afterwards he was ill. Every time he managed to shout, the men fell over, as if the sound of his voice affected them mortally.

I could get no more out of Antonio at this first meeting. He says he is expecting his brother today, and he is having an interview with a Colonel on Tuesday. So I suppose that if authority lets him give the rest of the story, he will. Radio and TV are after him, but he has accepted nothing yet, probably for the same reason, for he needs the money (which, so far as I know TV, is nothing or very little). On Monday he will again be examined by doctors.

Some of the results of my first interview are good, some are lacking. I deplore the fact that I did not try to interview the neighbours living around the field where the man was abducted. I deplore the fact that he discovered I was trying to tape-record (my memory is very bad) and so he became suspicious of me. But I am very pleased to have been able to find confirmation of the newspaper story in general, and to have obtained the psychologist's opinion, and to have met the man in his own surroundings and heard some of the story from him. I deplore the fact that I still do not have the means to carry out a thorough investigation when a case such as this occurs.

I do not believe a lie-detector test would be needed, the man's suffering and sincerity are obvious. And hypnosis might be resorted to at a later date. Prof. Silvio Lago is still abroad, so that will have to wait.

The interesting concurring details are those of the TV screen, which matched the 1968 (also September) Mendoza Peccinetti-Villegas UFO sighting, and the blood drawn from the middle finger (in that case it was their left hand). The fact of Antonio finding himself in a bell-jar corresponds to a description Onilson Patero made under hypnosis for Dr. Silvio Lago in '74, also to a very interesting investigation Mr. Cleto made and is still keeping under cover, which contains the twin image seen on the screen. Only in the case of Antonio La Rubia he sees himself in the near future in situations that actually came to pass.

This brings up the often commented matter of the space-time continuum change, and is remindful of the very strange Chilean events at the beginning of this year, where a sergeant was absent for 20 minutes and returned after approaching a UFO, with a 5-day beard and his watch marking five days ahead. There is food for thought here. The beings in this case were described fully by Dr. Neli, but I would rather await Antonio's description before relating it. He showed great enthusiasm and friendship towards me, and promised faithfully to come and see me and tell me all in the near future. I only hope he will be allowed to keep his word! I told him of other similar cases and this calmed him.

Antonio La Rubia is a man of his word. When I first met him, Oct. 8, at his in-laws' home, he promised to come and visit me at my home as soon as he was better, to give me a full account of his UFO experience. He came yesterday, the 18th, with his brother Arnaldo, unexpectedly, and my time for him was short, as I was about to leave home to give my classes at Cultura. Even so I managed to tape-record his narrative for about 40 minutes. He speaks fast, so there is a lot to transcribe, and what he says is full of meaningful content. I shall need more sessions with him to ask him a more complete set of questions, and this will take time.

He is now less nervous, less tense than he was ten days ago. He smokes less, still feels very thirsty and has lost weight. He is still under treatment for general recovery (vitamins, etc.). His manners are frank and simple, and he is polite. He says he still spends agitated nights, and needs a long rest. He has not dreamt of the beings, but sees them, and the scene of his experience, as he is still under shock.

Antonio said: "I get up every morning at 2:00, brush my teeth, clean up and leave at about 2:15 or 2:20. On that Thursday morning it must have been 2:20, for that is the hour that my watch stopped. I crossed over to the field near my house (the field is very large, not small, as '*O Dia*' said. I.G.) and, when I got to the corner, I stopped short, for there in the middle of it stood this huge thing, so large it went beyond the boundaries of the football field, which is more than 70 m. (228 ft.) across. I had not believed in flying saucers, but this was one, no doubt; it was shaped like a hat, a dull leaden colour."

At first Antonio had thought that it was the company bus coming nearer his home to fetch him, as the region was very deserted at that hour of the night and well known for its holdups. The UFO was resting on the ground, or so he thinks. He has gone back to the spot recently to see if it had left any traces but found neither marks nor charred grass. (Just as it was found by me. However, we might both have missed the marks. I.G.)

As soon as Antonio realized what it was he was seeing, he decided to run back home. That was his intention, but he was unable to do so, for at that very moment an intensely bright light was turned on that lit up the whole place. Antonio was standing by an electric post which became brightly illuminated with this deep blue light that struck him, and the whole electric post turned blue. Antonio commented that the light was fabulous . . . and at the same moment he noticed three 'robots' on the ground beside him. They reached above his shoulders, and he is 1.68 m. tall (5'6"), so we can figure they were about 1.40 m. (4'7"), or less, in height, except for their antennae, which jutted up out of the middle of their heads to over his height.

Antonio took the opportunity to go on describing their appearance to me. He repeated what he had told me 10 days ago, that their heads were shaped like 'American footballs,' and there was a band crossing the middle of this with what looked like a row of little mirrors, of two shades — one a little darker blue than the other.

He said their bodies were stocky, the trunk broader than his own (he is of muscular, but narrow build). They had appendages for arms, thicker than his own arms, which he compared to an elephant's trunk, because they narrowed down to tips, with no hands or fingers, or just one, if you count the tapered-off part as such, and it was as narrow as a normal finger. Their bodies were made of a rough substance forming scales. Asked whether this could be an armour clothing them, he said he did not think so, as their movements were free and easy, which would not have been the case if it were a covering. Their trunks were kind of rounded at the bottom, from which came out a single leg.

The first impression they gave, looking at them, was that they were sitting on something, but such was not the case. The single leg stood on what could be described as a platform, the shape and size of a saucer. He compared this leg and platform to the stools used on ships. All this outer part of their bodies looked like aluminum, but a dull shade.

In the field, one of the 'robots' stood in front of him, one at his side, and another at his back. He tried to get away, but after that blue light had been turned on, he found he couldn't move. He flung his arms around, but discovered he was a captive in what he called an 'invisible glass' and had described in his former meeting with me as a 'redoma' —a *bell-jar*. Otherwise he considered himself to be in a normal state, although very nervous. However, he could not move.

As for the beings, they floated along. They were all of the same stature and all exactly alike. Except that one of them was holding what looked like a syringe for shots. The being lifted its arm, pointing that thing at him, and he moved from where he was, without feeling it. In spite of himself, he was moved towards the 'redoma'. I asked him how he got into it, and he said he did not know, but for him it was because it was transparent. (This sounds fuddled, but he explained afterwards that it looked solid, leaden-like, from the outside, and was transparent from the inside.)

He does not know how he entered the 'redoma', but as he

approached it he felt a tremor and suddenly found himself in a corridor, white, of aluminum, and beyond there was a wall. Two 'robots' went one way, the other another way. He was inside the ship, and looked back down the corridor: he saw the square lying beneath (obviously the UFO had lifted from the ground). The view he saw was recognizable, and he said the vehicle must have been flying from south to north. He said the walls were transparent from inside out, so he could look out from all sides.

As he was looking back and out, a bright blue light came on again. He was now in a large circular hall, and the light, he believes, came down from the ceiling, where it was a deeper blue, then descended from there down to the walls where it became lighter until it melded with the aluminum colour of the walls. In this huge hall he saw a dozen men on one side, and another dozen on the other side, and they reminded him of school-children in a classroom, also because of their single legs that looked like seats.

It was then that Antonio finally managed to shout, for he had been struggling all the while, but hadn't been able to make a sound. Now he shouted: "What do you want? Who are you?" To his great surprise, he saw the little beings all tumbling over to the floor, just like little tin soldiers (the comparison is mine) and he realized the sound of his voice must have caused this. The light came on strong again, so strong that it blinded him, and after that he saw the beings getting up once more. He found he could hardly breathe; in fact, after the first tremor going into the ship, he had had difficulty in breathing. He fought for oxygen, and found there was none. He did not hear the sound of his own breathing, for he could hardly breathe.

As for the beings, I reminded him that he had told me he had heard them breathing. He confirmed this, for he was close to them, and heard what seemed like breathing, but they made no other noise. Everything was silent,

Antonio noticed that at the moment he had shouted, the beings had lifted their appendages to the tips of their antennae, holding them. The tip had a coffee-spoon appearance. Antonio lifted the little coffee-spoon he was using and said it was that shape, only duller looking, and the antenna came out from the side of the spoon. This tip rotated at an incredible speed, all the time, so fast his eyes could not follow. He said that while in the UFO at no time did he feel ill. He felt normal apart from his breathing difficulties.

After the light had come on again, he noticed there was a little, piano-like affair standing in front of him. Otherwise the great hall was empty — there was no furniture, no instruments, no panels of any kind to be seen anywhere. The piano-like affair was like a box, about 15 - 17 cm. (6-7 ins.) or so in width, standing on two poles so that it reached to the height of Antonio's chest. At the extremity of the box, on each side, there were antennae jutting up and to one side, the keys that had reminded him of a piano. There was also something that looked like a tin on it, where the beings put some little objects that they took from their belts.

Antonio now explained that they wore belts with a lot of what looked like syringes hanging from hooks of different sizes, around their belts; they took turns in introducing one of these things into the box, where he could not see, and each time this was done, an image appeared on the wall of the UFO, showing a different scene.

Antonio was shown a series of pictures in color. Every time this was done when the being had introduced the syringe-like thing into the box, it pressed a key and the picture appeared. (A complicated kind of slide-projector, it seems to me, I.G.) Antonio then told me he would try to remember the sequence of pictures as he had watched them.

1st: Himself naked, lying on an invisible (?) table, swinging his arms around, his legs together, and two of the beings examining him with their little bluish lights (coming from the syringe?) directing it at his chest, at his head, and another being examining his head with this blue light, that had no beam, Antonio observed. It turned everything blue, and even his hair was blue. When this scene was over, another being came over, introduced another 'thing' (slide?) and the 2nd picture appeared. Here he saw himself alone, still naked.

3rd: Antonio saw himself dressed, wearing his shoes, and with his carry-all for shopping. His teeth were chattering, he looked nervous, but no sound came from his lips, his arms swinging, without being able to leave. He corrected himself, saying: "No, I wasn't swinging both my arms, only one of them, the other was holding the bag."

4th: The next picture showed a horse and cart, the horse drawing the cart over a dirt road. He did not recognize where that was. There appeared what must have been the cart-man, ragged, a real peasant, wearing a straw hat, his feet naked, his shirt torn.

5th: He now watched the picture of a light orangey ball, and himself standing beside it.

6th: The ball appeared once more, this time bluish, a light blue, and one of the beings standing beside it.

7th: A dog was shown trying to get at one of those beings, also shown in the picture. The dog was big, and slobbering at the mouth, trying hard to get at the being, unable to reach it, and looked very angry. Then the dog gave out four or five barks. At this point, the being started to melt, from top to bottom, like porridge (mingao).

Antonio's voice took on an even more excited and puzzled tone, as he told me that at that moment one of the beings in the hall with him came out of the row, moved forward, stood beside him, and pointed one of those apparatuses at the dog, and the dog turned blue, and started to melt, exactly as the being had melted, and turned into porridge, or jam. It was shocking to see that happen, said Antonio. He described the 'ray' from the 'syringe', as different to the ones one sees in TV films, for it was, in fact, not a ray, but only a focus (a focussed light?) on the dog. He thinks the dog must have died.

8th: A factory, a factory of theirs, where the UFOs are manufactured. The scene was of incandescent white, and stretched along far out, so far he could not see the end of it. There were three rows. The two to the right showed UFOs that were nearly ready, and the row to the left had crafts in the making, at the skeleton stage. All three rows stretched out far out of sight. Everything was spotlessly white. There were 'millions' of beings 'robots'. They were just moving around, but he noticed no tools.

9th: This showed a train, like the Japanese trains now running in Brazil, but older, something the worse for wear, windowless, but just like our trains in other details, and Antonio believes it *was* one of our trains. It was entering a tunnel and was lost to view.

10th: This showed an avenue, like Avenida Presidente Vargas (one of the busiest thoroughfares in Rio) and it was jammed with cars. It was so jammed, the cars stood still.

Antonio's list seems to end here, but he must have been shown other pictures, as when he was shown the one where he saw himself dressed.

Antonio went on to tell me of when the beings drew blood from his finger. One of them came over to the centre of the hall where he was standing, took one of those 'syringes' from his belt, passed it over to his left appendage, where it started to rotate. Antonio says it spun so fast he could not follow it with his eyes. Then this thing was pointed at him, and his arm was lifted — against his will — and the syringe was stuck into the middle finger of his right hand. He saw the syringe filling until it nearly overflowed. He was sure it must have been his own blood, for it was the only color he saw in the whole place, all else was blue or white, metallic, and this was red. He could not see how it happened, for he had not felt the prick, and there was nothing to see after it was over. Then the being pointed at the picture on the wall, and drew (with it?) three circles, presumably with Antonio's own blood, and cut them through with an L-shaped mark.

It seems that the blood-drawing experience came before, or in between, the showing of the pictures. Antonio says that after the busy street-scene was shown he was 'thrown overboard', and fell into a street nearly opposite the Paciencia Station. So he had been tele-transported from near his home, three or four kilometres away to this place where they landed him.

Antonio says he does not remember how he found himself there, but he was standing on the road by the railway and wondering where he was. One of the beings was there with him; he looked back to discover his whereabouts, then up to the sky, and saw a huge 'balloon' going up. When he looked down again, the being was no longer to be seen.

All Antonio's belongings were with him, even his bag which had not been with him on the ship. Then he looked at his watch, and it had stopped at 2:20 a.m.

I asked him if he knew of anyone else who had observed the UFO that night, and he said there was. It was somebody who lives by the football field, but he is known as a drunkard, so he is not reliable. He went around saying he had seen a UFO on that same night.

Antonio crossed over to the station, and there was a man there, so he asked him the time, and it was either two fifty, or two fifty-five, so he put his watch straight. There was a bus passing at 3:10, and he caught it, and went to work, and even got there on time! He felt nervous and ill, his body aching all over. He drove the bus nonetheless, but now and again his vision darkened. He worked all day and the next, Friday, but when he got home that night he was done for, and went straight to bed. It was then that the reaction started, and he felt really ill.

At this point Antonio recalled that he had forgotten to tell me of one more picture, showing himself with smoke coming out of his back, and the pain and heat he now felt were as if this were happening. The picture in the UFO had depicted all the pains he was later to suffer, including his vomiting, etc. He told his wife nothing of all that happened to him. That Friday night his bowels were loose and he felt miserable. On the Saturday he was too ill to go to work, and missed Sunday as well.

That night the burning feeling started, spread all over his body, and was dreadfully painful. His wife rubbed some alcohol over him, and that lessened the pain, but it was throbbing. His head ached for days. On Monday morning, he went back to the bus company to say he had to quit, and he found difficulty in breathing and was burning and itching so much that he stripped and had a fellow there pour water on him. Even this did not alleviate him, and he started talking about the flying saucer. He still finds it difficult to take a deep breath, and feels an emptiness inside. His fellow workers told him he was looking as green as grass.

The burning affected him most where the light had been directed at him. I asked him how he felt when he walked, and he replied he felt an empty feeling, as if he were walking in a cloud, so I suggested "floating" and he said that was it, and he still feels that way, he feels light. This, 33 days after the occurrence!

The pain he felt from the burning that Monday was so great that it was as if he had been thrown in a pan of scalding water. The bus company nurse wanted to give him an injection to tranquilize him, and he refused, afraid that might make him worse, so they thought he had gone mad and even brought ropes to tie him down with. They called an ambulance and he was taken to hospital. They thought he was mad because he was babbling about flying saucers. He went to the Rocha Faria Hospital and finally was able to talk a little more quietly. The doctors gave him a hearing.

It was before being taken to hospital, however, that he got a hearing with the bus company psychologist, Dr. Neli Carbonell David who, seeing he was in great pain, transferred him to the bus company nurse, who called the ambulance. Dr. Neli, with whom I spoke again over the phone on Oct. 13, told me she had given Antonio complete clearance on her part, considering him normal.

The doctors at the hospital told Antonio there was nothing wrong with him, after giving him a general checkup, and he is surprised, feeling ill as he still does, that they tell him he is well. However, when one of the doctors visited him for his INPS (Workers' Employment Relief) and heard about his having seen a UFO, he called in six more doctors, saying the case was serious and worthy of study.

Although Antonio has not dreamed of the beings and his experience, he keeps reliving the events and seeing them.

This Paciencia case will require quite an amount of thinking about. It will certainly not please those who believe our visitors are only beautiful goodie-goodies come over just to save mankind. I do not exclude those either, believing as I do that the universe follows the same, or similar pattern everywhere, so good and evil, or bad, are balanced in all parts.

We have not yet had the time to evaluate how great the harm wrought on Antonio, one more extra-terrestrial guinea-pig, has been. Time will tell. But we are ready to criticize the fact that an honest, hardworking husband and father should be made to face an attack from the unknown, and nobody, where it matters, does a thing about it. The argument that misfortune comes in many ways, and he could have been assaulted, or murdered by robbers, or killed in an accident, does not stand. These are everyday occurrences that can happen and we can deal with them.

But Antonio is one more victim of a series of abductions, the majority of which remain secret and unknown, either because the victim does not return, or because he will never talk about his experience. Typically, Antonio exclaimed: "If it hadn't been for the pain, which was excruciating, I would never upon my life have told anyone of what had happened to me!"

Antonio was lucky in finding a psychologist right from the start, and to have a good reputation. In the Barra do Pirai case that I investigated two years ago, where two working women were exposed to the light from a UFO, and one of them died months later of pains and aches that were never properly diagnosed, and the other suffered after-effects for over a year (I do not know of her present state of health), nothing was done. Prior to their experience, they had been healthy.

The similarities in Antonio La Rubia's case, compared with others elsewhere, are many. The first that came to my mind was the Mendoza, Peccinetti-Villegas UFO encounter, because of the blood being drawn from their middle finger (of the other hand) and because of the TV screen images. TV screen images were shown inside a UFO in the Bellizzi, 1969 sightings that I investigated too. The floating feeling subsequent to a UFO encounter is characteristic of many cases, including the Jonil Vieira case, 1967. The appearance of the beings, their scaly outside (Antonio assures me it must have been their skin, not a covering) reminds one of Pascagoula and others. There is a 'family resemblance' between the Pascagoula drawing of the being and the one Antonio made himself.

But the most puzzling new facts in this case are the showing of the sequence of pictures, not in themselves as such, but *what did the beings wish to communicate?* This is the task for many specialized scientists to try to unravel. What message did they wish to convey? The simplest seems that, as we harm them, they can harm us. That there are many of them, as there are many of us. That they can tell our future, but we cannot tell theirs. They isolate people in an invisible bell-jar. And so on.

Shall we have time to reflect on all this before they come over in larger hordes?

‘Psychic Photographer’

Dorothy Izatt, the psychic photographer.

As communication with UFO beings apparently is conducted at times on a psychic level, humans with psychic gifts are presumably best equipped to function as media for these visitors. One whose camera work suggests this is indeed the case is Dorothy Izatt of Richmond, B.C. Without any special facilities or training, she has produced images so strange yet containing such definable forms, they give one the feeling of peering into another world, which perhaps is true. Although she has readily allowed experts to examine her pictures and equipment (she favors a simple Keystone XL movie camera and sends all her film out for processing), the mystery remains unsolved.

It may well be that the secret is within Dorothy herself, for her psychic aptitude is pronounced, a trait that she and her six sisters seem to have inherited from their father. She remembers that as a girl in Hong Kong, where they grew up, she could see vision-like pictures on the walls of her room.

Now a young-looking mother and grandmother, she is far removed from any tangible association with the occult, yet her unusual qualities persist perhaps more strongly than ever. For instance, there is the wide spectrum of her sight. While she can see the auras of living things, as can others of her kind, there is more to it than that. After she and her husband had a micro-wave oven installed in their kitchen, she remarked on the beauty of the colors it sent out when the heat was turned on. Only then did she learn that the micro-waves were invisible to others!

In taking her pictures she may be prompted by something she

glimpses in the sky (occasionally also seen by others) or by a sudden feeling of warmth on her face or simply by an inner urge to do so. But whatever the case, her camera somehow ranges far beyond the city limits of her home. It would seem she has "friends" out there who sense she is a channel through which they can express themselves photographically.

Many of Dorothy's best shots, usually taken from her lawn or open window, appear on a single frame of movie film. Here are examples (all under copyright).

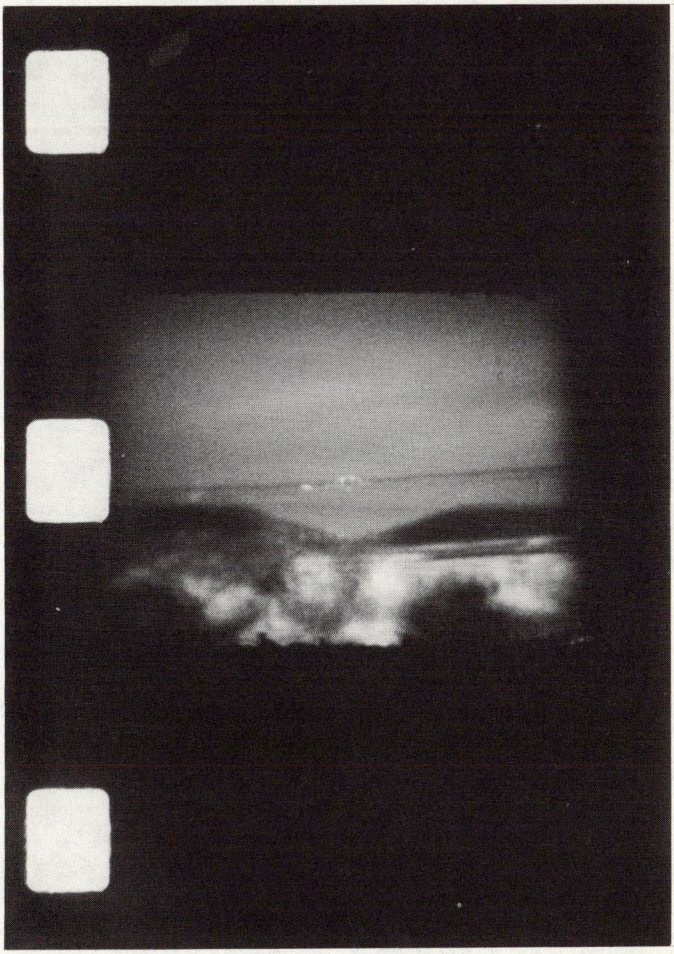

This picture was taken outside Dorothy's Richmond, B.C., Canada home which is miles from water.

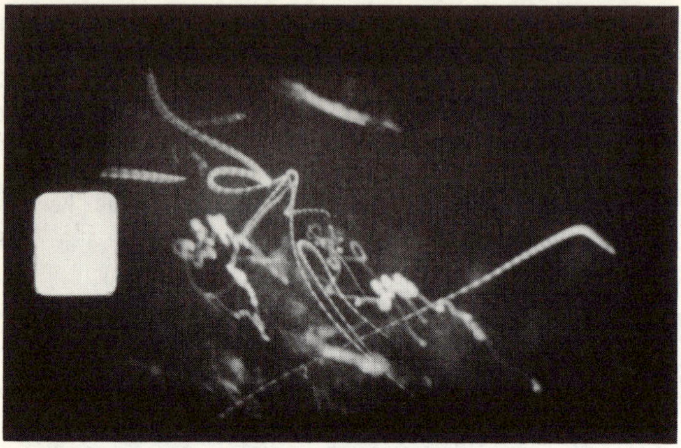

This picture was taken in the evening from a motel in Washington. The picture shows a white object with green, blue, gold and orange lights.

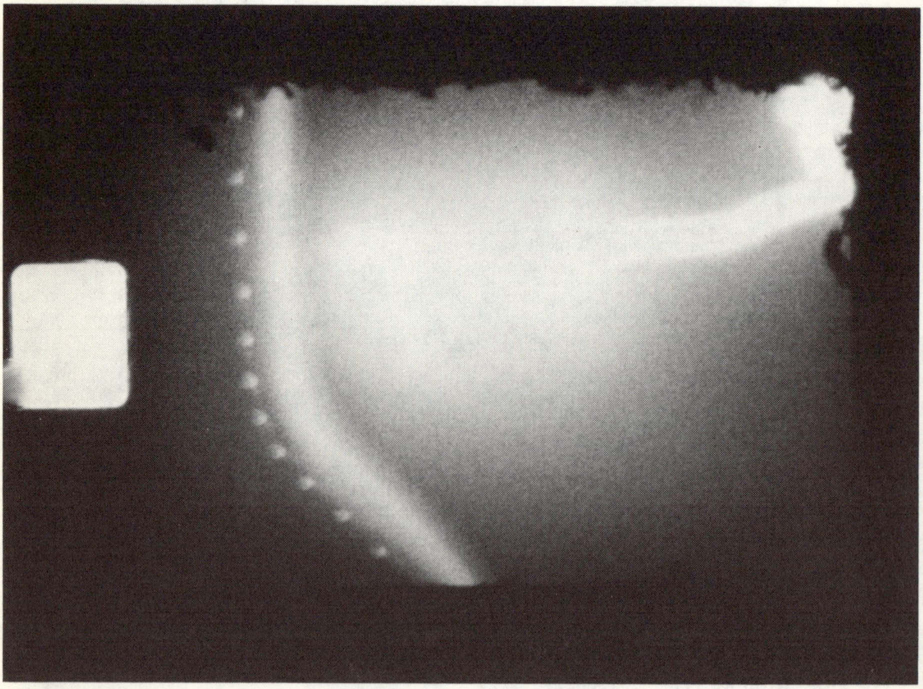

This photo is of a section of a "huge object" Dorothy saw above her house. The brilliant glow suggests a power discharge.

Perhaps the most significant photo revealed a helmeted figure as it peered out between the windows of the craft overhead. The hook at bottom right seems to be attached to a small bright object.

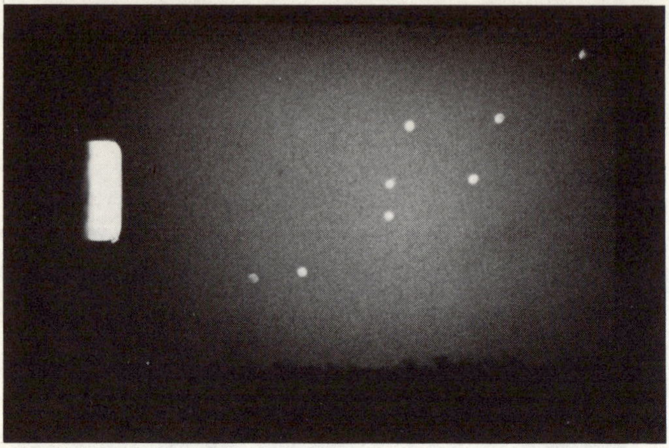

The weather was clear when this picture was taken from a window facing south. Dorothy saw a group of round white objects covered with red and blue lights. During filming, they each took off in a different direction. A few were captured on film.

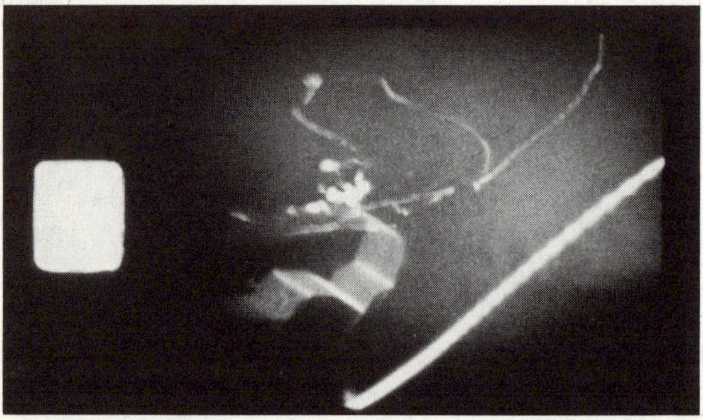

This photo was taken as Dorothy looked at a supposedly empty sky.

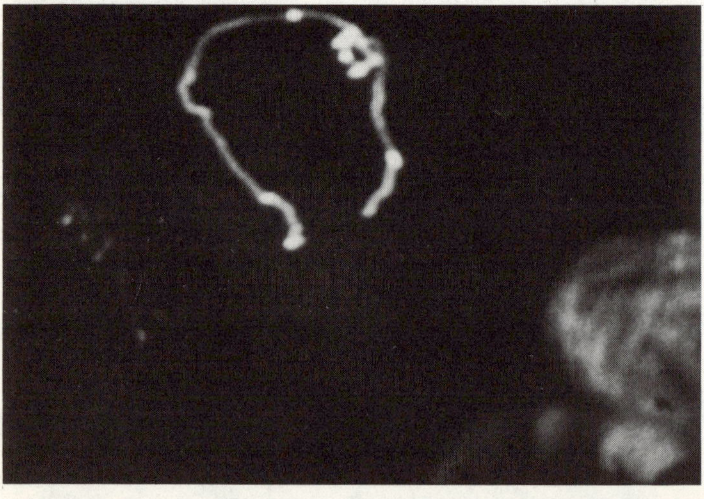

Dorothy said this photo showed the railing of a craft she saw overhead. A figure appeared to stare out from the top of the craft.

A small gold ball moving almost quicker than the eye traced this horseshoe pattern. The "blob" appeared nearby.

Chapter Five: Looking For Power

Like the dream of some Paul Bunyan-type trapeze artist, a vast high-wire complex of power lines borders the St. Lawrence river and enters Canada's industrial heartland. Since our UFO visitors gravitate to electricity, this network of lines is evidently like a dream for them, too, and sometimes the result is trouble.

Because of what is generally thought to be their electromagnetic effect disrupting the flow of energy, UFOs can create havoc when in the vicinity of a power supply, a problem that has been a special subject for ufologist Claude MacDuff of Montreal. When a series of blackouts in Quebec a few years ago coincided with a series of sightings, he saw a chance to draw conclusions.

"The apparent relationship between UFO sightings and power blackouts was evident in Quebec province during July 18-24, 1971," he advised us while preparing a book on the subject. "In that period of one week a succession of power failures occurred, corresponding with UFO observations in the same areas. Some of the objects were observed directly above power stations of Hydro-Québec. Twice in this same week the Montreal area was hit by blackouts . . .

"A look at the map where objects were seen shows that most of the sightings occurred close to the St. Lawrence river. This ties in with the pattern in other countries where UFO corridors are known to follow the main waterways."

With this picture in mind of UFOs operating where power is abundant, let's look at the following sighting report made to us from the same area by Wido Hoville who was soon to join MacDuff and others in publishing the periodical *UFO — Quebec* (which carried a more detailed account of the incident in its first issue):

> Scene of the sighting was St. Cyrille de Wendover, Quebec, which is three miles northeast of Drummondville where several textile companies are located. Immediately to the north are two 730,000 volt powerlines. The town lies close to the railway line between Montreal and Quebec City and to the Trans-Canada highway.
>
> Mr. and Mrs. L., whose name is known to me, came home from a holiday in Florida on the evening of June 22, 1974. Their home is one of nine concrete based trailers in one section of St. Cyrille, though there were five at the time and the witnesses were the only residents on the street.
>
> The two are 26 years of age. When the incident occurred, Mr. L. was watching the late show on TV, while his wife, tired from the trip, was already asleep in the adjoining bedroom. Shortly after the end of the show at 1:15 a.m. Mr. L. was preparing to go to bed

when he heard a strange sound outside. It was some sort of boom. . .boom. . .boom as if something had fallen on the ground. Lifting the curtain of his living room window, Mr. L. saw a reddish-orange circular object hovering over a field on the northeast side of his trailer.

The witness went into the bedroom to wake his wife and while there he heard a buzzing sound. Looking from the bedroom window, he saw a figure that had apparently emerged from the object and was hovering not more than 15 feet from the window. Shocked, Mr. L. went with his wife into the living room from where they observed three more "robots" as they called them because the strange creatures looked "stiff." The figures were close to the trailer next door, examining the shaft and wheel assembly. (This trailer had just arrived and was not yet installed on its foundation. Sand was spread around as base for a lawn.)

From then on the witnesses looked out only occasionally to see if the "robots" were still there. They were undecided what to do and were badly afraid. (It was said that an unidentified woman also saw the figures and called the police, who did not come.) At one time during their observation which lasted on and off until 4:20 a.m., Mr. and Mrs. L. saw about 15 "robots" together in a line close to a nearby creek. They stood there for over five minutes and moved suddenly together as if on command.

The "robots" were about six feet tall and had neon-like "infra red" bars across their bodies which switched off and on. The figures did not seem to touch the ground and no traces were found in the sand next day.

More than any item I can think of, that one by Hoville (who went on to say the witnesses did not see the figures leave) explains what this and the following chapter are about, which is different aspects of our UFO visitors' regard for electrical power. While their continual and wide-scale involvement with electricity does not seem to be a purpose unto itself, it is very apparently an essential means to achieve their general range of purposes whatever that may be. In short, without the use of electricity at their immediate command, which has to be considered an unlikely predicament for these advanced creatures, it seems probable they would find Earth an inhospitable place not worth bothering about.

In Hoville's report, note first the proximity of powerlines. As we shall see, there are cases in which the point of attention for a UFO is so obviously a cable of this sort, or something else connected with the transmission of power, it is probably no coincidence that the flow of electricity somewhere enters the scene of many sightings of a more general nature.

Next note that a booming sound was heard at the time of sighting. This is considerably more rare in a UFO encounter than the presence of electrical apparatus, but at the same time it provides a deeper insight into what our visitors' concern with devices of power may be all about. A ufologist in New Zealand, Bruce Cathie, has developed some remarkable ideas about the reasons for their concern which we will consider in the next chapter.

Finally note the presence of "robots" on the scene. This is a rare item indeed in a sighting with such electrical implications, particularly as the figures themselves, with their infra red glow, appear to have been charged up to do whatever they had to do. Unfortunately, in the brief moment of observation, their actions did not mean much to the witnesses. But I do have on file a case in which electrically-oriented robots showed much more purpose, as well as a case in which the occupant of a UFO near a power installation seemed to have a human form. Both these cases, which we will look at in the next chapter, emphasize the extent to which our visitors will go to make contact with electrical apparatus.

Now, turning to the first point in Hoville's report, when he mentioned that the sighting in question occurred near powerlines, he brought to mind this report I wrote for CUFOR:

> It was 10 o'clock on the hot still night of August 26, 1967, when at least a dozen residents at McCleese Lake on the Cariboo Highway in British Columbia saw a glowing blood-red object about the size of a soccer ball move into sight from the south. It was travelling directly above newly installed cables leading to the giant Peace River power project-to-be far to the north.
>
> "It didn't seem to be in any hurry and it stayed right over those lines as if it was studying them," Mrs. Alfred Beck told me.
>
> A noteworthy point of this incident was that the Peace River project had not then started operation, so no power was passing through the cables at the time. Nor was the ball of light connected to the lines. It was well above them.

In the years that followed, with power flowing through 700 miles of cables from Bennett Dam in the north to Vancouver Island at the southern end, the visitors' interest became so pronounced that at various times sightings were reported along the full stretch of the system.

While it seemed remarkable at the time of the project's installation that a strange aerial object (and no doubt others that went unseen or unreported over the sparsely populated country) appeared as if from nowhere to examine the empty lines, I should not have been so surprised. After all, my records showed there had been earlier sightings above the older Canadian National Telecommuncations system running beside the Alaska Highway, and that system crossed the powerlines from Bennett Dam like the bar of a T. It should have been obvious to me, then, that UFOs would explore the new southward extension out of curiosity, if nothing else.

Since my mistake may have been in underestimating the prior UFO presence along the northern highway, let's look at a sample of the action on that road in 1965, which was before the powerlines going south were installed. The scene was Swift River, site of one of the CNT micro-wave towers, where my wife Lillian and I stopped for a while in 1968 during our first of several Yukon visits.

Like many others in the Yukon who were impressed by the number of sightings made in their part of the country, Mrs. John McCreedy of Swift River often scanned the sky at night in hope of seeing the unusual. She was rewarded several times — "at least half a dozen" as she recalled — by spotting objects which did not match the description of any aircraft known to her. And that covered a wide field, for Mrs. McCreedy, wife of the public works foreman at Swift River (now deceased), was a member of the Ground Observation Corps until it was discontinued. In that capacity she reported and described any aircraft that was off the designated flying routes.

"The objects had the size and appearance of satellites which we often see overhead up here," Mrs. McCreedy told us. "But they acted so differently I am positive they were not satellites. They moved far more quickly, or slowly."

Up to that point, however, the McCreedy family had no reason to think that the CNT system might be the cause of the aerial visitors' interest beyond the fact that all the objects seemed to follow a ridge of the Cassiar mountains running parallel to the route of the micro-wave towers. Then one evening Marina McCreedy, aged 13, while outside with other children saw a flying star-like object that seemed to focus on the main attraction.

"It moved in the direction of that micro-wave tower over there, or maybe it was closer to that round hill just this side of it," she told us, pointing to slopes in the distance. "Anyway, when it got there it seemed to slow down a bit and suddenly it sort of shot off flashes or sparks. Then it went on again."

(With each transmission tower in the system fed from below by a cable carrying 10,000 volts of electricity, there appeared to be only one explanation at the time for this behaviour. Evidently it was another case of our visitors expressing their interest in electrical power. But 10 years later, in a manner typical of UFO affairs, there was reason to doubt this tidy answer, for a massive deposit of tin was discovered later in that round-topped hill which today is being developed into Canada's only tin mine. Was it that deposit rather than the nearby CNT tower that caused the UFO display? We will never know but, in view of the UFO presence along the entire CNT system, a reasonable guess is that the electrical output attracted our visitors in the first place while the tin deposit later prompted this special show of attention.)

The vicinity of that round hill was the area of attention again when it became the scene of two sightings within a few nights of each other by Mrs. Heidi Radford, whose husband Peter was a diesel maintenance supervisor for CNT. He introduced me to others who had sightings near the system.

Speaking of the first experience, Heidi said that as she was returning from a curling game one clear, cold February night in 1965, she saw a reddish-orange glow above the round-topped hill.

"I saw it for a second or two before all at once it seemed to burst," she said. "Flashes went off in all directions and the thing was gone."

This remarkable display became more mystifying when, in almost identical conditions, she had her next sighting of a similar glow, again over the round hill.

"This one darted about and made me think it was looking for something," she said. "It moved so fast it seemed to be in different places almost in the same second, but it never went far from the hill until suddenly it disappeared."

Could this have been a case of one UFO searching for another?

The mystery of what happened up there so intrigued the people of Swift River that a search party was organized to investigate, but nothing unusual was found, which raised once again the long unanswered question of what happens to UFO wreckage? Do the aliens have some instant self-destruct device that atomizes a craft whenever disaster is certain?

Luckily for me questions like that are not what this book is about. But since this study is international in scope, we can move a step farther here with a look at a reconnoitering UFO in Brazil as it appeared to two technicians on the site of a microwave tower similar to that at Swift River. This report was another by Irene Granchi, our valued correspondent in Rio de Janeiro.

After explaining the incident had occurred near a microwave tower outside the town of Barra do Pirai in the late afternoon of February 1, 1976, and that the two men saw the object as they prepared to leave in their car, Mme. Granchi wrote:

"It was spinning, size comparable to that of the moon, and its light was reflected by the low clouds above it. The day was still cloudy, and it had rained earlier. The UFO's true size the men decided, must have been that of a car or a helicopter. It was soundless. They tried figuring out its shape, unsuccessfully, because of the brightness of the light surrounding it. They stayed on for half an hour, and finally returned to the tower, only to find the light in the cabin out, and neither of them had turned it off. They did not wait to see if the UFO had left, and went away.

"Jorge (Pedroso de Lima, one of the witnesses,) told me of other events involving the micro-wave towers. He said that the year before an employee named Itamar was caretaker at the Barra do Pirai tower and nightwatcher combined. This man rang the CTB (telephone company) offices and the Barra do Pirai police on several occasions, in a panic, saying there were prowlers on the premises, and he had locked himself in. This occurred whenever lights appeared in the sky above the tower, causing the lights on the tower to go out. After Itamar's calls, when the police went up to check, they found nothing. But Itamar always found circles of charred grass on the ground surrounding the tower, after each experience. He was finally fired by the company, and there has been no one to replace him.

"Jorge de Lima took me and Mr. Tinoco, the Vice-Mayor of Barra do Pirai, to visit the Barra do Pirai tower, where I took photos of the site. On our return down to the town, we stopped at Itamar's house, and spoke to his wife, for he was out. She confirmed her husband's experiences, saying he had often complained and come home panic-stricken. She herself had often seen lights travelling in the direction of the tower, and so did her children, who were with her."

Now, to re-cap for a moment, it will be remembered that in his Quebec report Wido Hoville said first of all that the incident he was writing about occurred near powerlines. He then said a booming sound was heard. We will consider that sound effect next but while we are discussing our visitors' interest in power transmission whether used in communications or for electrical supply, let's look at two examples from our files which make that interest even more obvious.

One was provided by Walter H. Andrus, Jr., international director of MUFON, who investigated a case in March, 1973 that involved a well-observed UFO making a close-up survey of a power station at Grand Tower, Illinois. Having interviewed Oscar Wills, 57, who was on duty as shift engineer the night the object was seen, as well as another employee and his son who first saw the UFO from their home, Andrus sent us a report of the result. Helpfully he included a copy of the paper published by the operators, the Central Illinois Public Service Company, which also reported the sighting.

Andrus wrote, in part:

"Oscar Wills described the object as a ring of lighted panels or windows with definite spacing between windows, each emitting pulsating light varying from a dark red to orange to white sequenced in a clockwise rotation like a theater marquee. The pulsations gave the impression that the object was spinning clockwise, but the window separators remained stationary. Oscar said the object was doughnut shaped but more like a wedding band, since only the lighted panels were visible. He got the impression that if the object was a solid structure the bottom was concave, since nothing was visible illuminated from the outer lights as he observed it directly over his head on two occasions.

"Oscar walked to within 150 yards of the transformer bank observing the object with the greatest curiosity from the steep angle. He became apprehensive and a little afraid, so decided to go back to the power plant and ask the two other employees on duty to come out and witness this phenomenon. As he turned to go back to the plant the object which had been hovering directly above the transformer bank started moving directly toward him, passing over his head.

"As the object darted noiselessly past the northwest corner of the building, he lost sight of it for a moment when the plant building obstructed his viewing area. He walked rapidly around to the west side of the plant that faces the Mississippi River and noted that the object

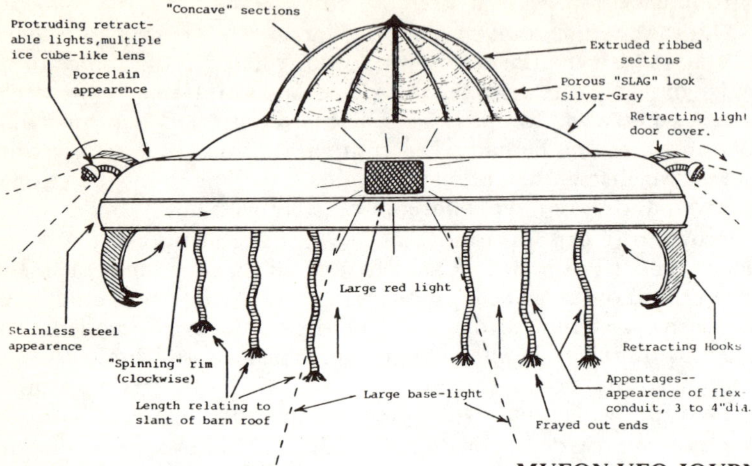

MUFON UFO JOURNAL

was now hovering directly over the middle (No. 2) of the three water pumping station buildings associated with the plant for water circulation purposes. Oscar then walked up the ramp to the pump station house and on to the sea wall protecting the plant from the Mississippi River. He was now directly under the hovering object. He said it would take a circular object 4½ to 5 inches in diameter held at arm's length to cover the object at this point. An earlier estimate of the object at 25 feet in diameter was made when it was hovering over the transformer bank.

"After having observed the object in both locations for 7 to 8 minutes, he was called back by the public address system operator to the telephone within the plant. It was Willis Hughes, (first witness) who wanted to know what Oscar could see from his vantage point. Oscar cut the conversation short so that he and the two other men on duty could go outside to view the object. When Oscar went out the west door of the plant near the pumping stations, the lighted object was nowhere in sight, so he had been the only witness at this favorable location. Both the switchboard and turbine operators went outside to look at it this time.

"When questioned about any sounds associated with the unusual object, Oscar stated that there was a slight hum originating from the high voltage (66,000 volt) transformer bank and that sounds emanating from the power plant were insignificant. The rushing of the high water of the Mississippi River along the adjoining sea wall could be detected in the quietness of the night. Based upon the detectable and identifiable sounds, there was no apparent noise created by the craft overhead. . .

"Central Illinois Public Service Company engineers from Springfield, Ill. immediately consulted all of the automatically recorded charts from each generator at the plant to determine if there were any voltage or output fluctuations at the time of the observation. They found none and no power outages were reported from the areas that this particular plant serves in southern Illinois. Oscar confirms this fact . . ."

While that episode, occurring as it did in the center of a generating system, left no doubt that electricity was the name of the game, the physical involvement of electricity was implied rather than observed. In the next case, recorded in California, we go well beyond that. Here the scene literally glowed with the missing item.

Although coverage of this outstanding incident by three major UFO organizations (APRO, MUFON and CUFOS) and the *National Enquirer* produced abundant detail, some of which we will look at in a moment, an earlier item in a local newspaper gave a good outline of what happened. It said:

> It was a nightmarish experience I'll never forget. I didn't believe in UFOs before, but I do now. You'd have to see it to believe it, but so help me what I saw — and felt — about 1 a.m. Friday (Sept. 10, 1976) outside our house is the truth.

That's what Bill Pecha, Jr., who does machine and bankout wagon work in his shop at 1650 Wilson Avenue, Colusa, told the *Sun-Herald* in describing what happened after the power went off (the power outage lasted for about nine minutes) while he was watching a Channel 12 television program. His wife and two children were asleep at the time. Said Pecha:

> When the power went off I started outside and saw this huge glowing thing hovering over our house. It looked to me like it was 150 feet across and from 10 to 18 feet in height, shooting out beams of bluish light. It looked like a huge inverted cup and saucer with the outer brim spinning clockwise and an inner disc spinning counter clockwise. When I got outside I felt funny. Seemed like all the hair on my body was being pulled up.
>
> There were three of these things, the big one and two smaller ones on the high power line in back of the 'Slim' Davis house. There was no noise, no sounds of any kind — just a huge blue halo of light with the big UFO maneuvering about at amazing speed. I awakened my wife and she and my son and daughter saw the whole thing, too. So did Les and Gayle Arant because I awakened them as I drove into town to report the UFO sighting to the sheriff's office. At that time this 'thing' was heading toward the fairgrounds and disappeared in a southeasterly direction, like toward Sacramento.

Pecha said the Arants drove into town, too, as a report was made to the sheriff's office which apparently had received other reports from Colusa area residents who had seen the strange glowing object.

Pecha said he was so frightened that he drove into town as fast as he could in his pickup with the lights out. But, he said, the bluish light from this thing, about 50 feet off the ground, seemingly followed him, then disappeared. He further described the events which transpired thusly:

Slim Davis and his wife were asleep. There was a bluish light all over their place, and coming down these shafts of light it turned to an all-white color. This thing darted about, clear out to the foothills and back. On each side it had two things about eight feet long which hung down and curled around underneath it. They looked like wire bristles. The power line poles were glowing and the two smaller UFOs each had lights.

All of this happened, I guess, in a matter of 15 or 20 minutes.

Although the later inquiry established that this newspaper account was not completely accurate, the investigators came away convinced that a remarkable incident had occurred basically as Bill Pecha described it. Without being outside the house to have a good look at the object and more concerned with her husband's excited behaviour, Mrs. Pecha nevertheless saw a brilliant yellowish light with a white beam shining from the bottom approach from the western foothills. Next she saw it illuminate the roof of their neighbors' house. To quote the APRO *Bulletin* (vol. 25, no. 9):

" 'Well, then,' Mrs. Pecha recounted, 'with this screamin' and what I saw out there, I knew I better move. I didn't — to me I still did not have any idea what I was lookin' at. I — I really didn't know what it was . . . I was lookin' at this light and it was scarin' me. But I could not imagine how come it was scaring Bill so much more!' "

To judge from his own words, Bill had good reason to be scared. At one point outside his house, he said, he was almost directly under the object which, flying just above his TV antenna at a height of about 50 feet, had six wavy cables with frayed ends hanging freakishly below. At opposite sides of the circular vehicle, later estimated to be about 90 feet in diameter, were two hook-like appendages which swivelled up while the cables were drawn in as the craft moved off. In the center a shaft of light beamed down, ending in mid-air in true UFO fashion.

Just as startling as all this was the effect on the atmosphere around the house. It seemed literally supercharged. Pecha said that when he went outside in his shorts, the hair on his arms and chest stood on end, crackling and snapping, and was topped by tiny haloes of blue light. The ground under his bare feet also seemed charged with electricity.

Back indoors to rouse his family, Pecha saw more of the weird show through a bedroom window. While the first UFO streaked off to the foothills about 20 miles away and returned in seconds, two other objects about half the size appeared in the distance and aimed blue light-beams at the high-tension cables running in that direction, causing them to glow bright red. Pecha said arcs of light shot through the beams from the lines to the craft, as if they were "zapping power."

(Power officials pointed out later that these lines were part of a federal system and had nothing to do with the blackout which simultaneously hit the area. That was caused when a sub-station on the local system blew an overload transformer, for reasons unknown.)

Terrified that invaders were after them, Pecha rushed his wife and two children into his camper truck and headed for town, not to see the sheriff as the newspaper said, but to make contact first with their friends, the Arants. It was a good choice as it immediately produced two more witnesses. When Mrs. Arant came to her front door in answer to Bill Pecha's frantic pounding, she promptly saw the aerial light with a dome on top to which he pointed. Following quickly, her husband also saw the object.

Pecha phoned the sheriff's office from the Arants' house, and when two policemen arrived to hear his account another strange note was added to this night of high tension. One of the officers said Pecha was so upset, he touched his arm to gain his attention. As he did so, he felt a strong electric shock up to his elbow.

Chapter Six: Earth As A Power Base

"Aerial" in Antarctic seabed as photographed by USNS *Eltanin* — ICUFON Archives.

From APRO *Bulletin*, January, 1978:

Beginning in early December the Eastern Seaboard of the United States experienced a series of atmospheric "explosions" or "booms." At 10 a.m. on December 2, South Carolina's shores were shattered by a blast so loud scientists said it would have taken the equivalent of 100 tons of dynamite to duplicate it. At 3:45 p.m., a few hundred miles north, a similar blast occurred, on the New Jersey shore, toppling knick-knacks off shelves and causing earthquake meters 50 miles inland to jump wildly.

But there was no earthquake, and an airplane pilot reported seeing a flash of green light in the afternoon sky.

The next blast was heard and felt by people in Connecticut on December 20. Then on December 21 and 22, New Jersey experienced the phenomenon again.

Less than a month later, on January 12, walls rocked and windows were broken when a blast shook the Charleston, South Carolina area.

The most recent blast took place on the night of January 19

or 20 (exact date not known) off New Jersey. A TWA pilot reported seeing a blue flash in the air. . .followed by an orange flash on the ground and accompanied by a power blackout over a two square mile area which lasted a few seconds.

Although several theories have been offered (sonic booms from aircraft, meteor showers, thunderclaps), no one has come up with a suitable answer to the puzzle . . .

It has been learned that similar unexplained blasts were heard and felt in 1890 in India, Australia, Scotland, New York and Connecticut.

With that news item we turn to the second of the three points taken from Wido Hoville's report of an incident in Quebec. This one concerns the booming noise heard by the witnesses as if some great weight had fallen on the ground . . . or was it *into* the ground? As the APRO *Bulletin* recalled, booming noises of a much greater magnitude were heard off the east coast in the winter of 1978-79 and created a mystery which is still unsolved, just as in its own lesser way the Quebec case has gone unanswered. Similar booms heard off the Oregon coast in 1982 are also unexplained.

Perhaps of major related importance in the APRO report is that bit about an aircraft pilot seeing an orange flash on the ground. Was the flash caused by something, maybe a bolt of energy in a form known only to our visitors, being shot with explosive force from the air to penetrate well below the earth's surface? If so, was the same sort of operation going on at sea which would account for the series of booms offshore? Did our visitors have a major building project underway to improve the efficiency of their global grid?

While that may seem like reaching too far into science fiction to find answers, Bruce Cathie, a commercial pilot in New Zealand and author of books on ufology, has developed what many consider a plausible case for his theory that space beings make use of Earth as a power base for their operations. He believes that over the centuries they have systematically plunged beacons into the ground to form a grid which serves as a navigational aid and also supplies their craft with power. Presumably this would be in addition to whatever they obtain from other sources such as those described in Chapter Five.

Cathie's ideas first won attention down under in 1967 when he was able to show that four well publicized sightings on New Zealand's coastline fitted neatly into lines on his grid drawn two years earlier. He then explained to *Copley News Service* how he had plotted the grids:

> I recognized the fact of UFOs' existence when, in 1952, together with six friends, I witnessed a huge ball of light, circled by a smaller red light, over Auckland. Other independent reports came in on the same UFO. As a pilot I felt that further opportunities for UFO observation would arise.
>
> In 1956 I saw a UFO over the top of the South Island where many similar sightings had been reported since 1909. In March of

1965 I saw a USO — an unidentified submarine object — in a northern harbor. It was cigar-shaped, about 100 feet long and 15 feet wide, and it was in about 30 feet of water. It was not a submarine.

The South Island positions and the USO position gave me definite fixed positions. From then on I began to keep a careful record of all verified sightings.

On my map I extended the line north and south from the USO position, to find that the line intersected points where other objects had been reported. In a similar manner a grid of lines at right angles to this began to emerge, spaced regularly at 30 minutes of arc.

The grid proved its accuracy with every new, verified sighting — all fitted onto one line or another. I became convinced that what I had found in this pattern was a grid, representing regular flight lines. The best evidence came with the four confirmed sightings over a relatively small area at Hawkes Bay.

Not content with a grid covering only New Zealand, Cathie sought some means of extending the pattern to cover the whole earth. For this he needed a reference point to give him a clue for the global pattern.

The reference he needed appeared in a photograph taken on August 29, 1964, by the American survey ship *Eltanin,* during a routine sweep off the South American coast. On a frame of film exposed over a barren sea bed 13,500 feet down was an antenna-like object which appeared to be metallic, with six cross-bars ending in knob-like protuberances.

What was it, and how did it get there? There were no answers forthcoming and the photograph has been forgotten. But Cathie had a hunch about it.

From the position of the object on the globe he measured a point on the exact opposite side of the earth, which turned out to be in Siberia. This line gave him the basis for a global grid system, into which his existing New Zealand grid fitted perfectly.

Now Cathie plotted UFO positions from all around the world, and these, together with complicated harmonic math, were the pattern for his global grid, which proved to have an equator identical to the earth's magnetic equator.

"In fact there are two interlocked grid patterns," Cathie explained, "and possibly there is a third system. The math I used always relates accurately and predictably to earth's mass, gravity acceleration and the speed of light."

What is the grid, then; what does it actually represent — supposing Cathie's conclusions are valid?

"I believe it is a navigational and power system radiating energy," Cathie declared. "It suggests to me that the earth is a power base for the UFOs, and I believe that similar systems will be found on the moon and Mars.

"I believe that the grid has existed for centuries, that it fell into disrepair for whatever reasons, and that it is now being rebuilt.

"I believe that there is, quite literally at our feet, a fantastic source of power that will put our space science, our power generation systems and our transportation and communications industries into obsolescence."

In 1968, soon after that interview Cathie had his first book published. Entitled *Harmonic 33*, it included a chapter describing what may have been the actual moment when a new power capsule or beacon was driven into New Zealand soil. The chapter opened with this report from the Auckland *Star*:

> Thousands of Aucklanders were awakened at 4 a.m. today by a massive explosion which tore apart a foundry at Rosebank Road, Avondale, and caused wide-spread damage to adjoining buildings.
>
> No one was injured in the explosion. But one family, whose house, about 60 yards from the blast, was lifted from its foundations, said everything "lit up like day" as glass showered into the house from blown windows.
>
> The blast, which was heard as far as Waiuku, 27 miles away, ripped 30 ft. long sheets of roofing iron from their supports, and tossed them into twisted heaps on neighboring gardens up to 100 yards away. It snapped heavy steel reinforcing beams supporting the foundry roof, tore concrete blocks from the foundry walls and threw them into the gardens.

The report was followed by an interview with E.C. Knibbs, owner of the house next door, during which it was learned that immediately before the explosion Knibbs' son was awakened by an unusual noise. He described it as a "loud hissing sound." (This calls to mind that in his report from Quebec Wido Hoville said a "buzzing" sound was heard as a circular object hovered over a nearby field. It suggests that a flying object, though unseen, may have been involved in this case as well.)

Cathie wrote that when he saw the news item he thought of UFOs at work on their power project and went out to investigate. He found several significant points. One was that a small crater-like hole had been punched into the concrete floor of the foundry and that across this was a circular marking about nine inches in width. Cause of this could not be explained any more than the rest of the accident and Cathie suspected the hole marked the spot where the aerial had entered the ground.

Also he learned of the strange experience of the residents next door.

"Mr. and Mrs. Knibbs said they were first awakened by a muffled crump, like a bomb explosion in the distance, followed by an extremely loud crack like a monster firework. The house shook and rolled, and all the windows blew in, showering the house with glass.

"After the initial shock they climbed out of bed and peered through the window which overlooked the factory next door. They found that most of it had disintegrated and they could see right through the wrecked structure.

"The sight that confronted them was most unusual. A thin pencil of light extended vertically from the floor level of the factory where the hole was later found, to about roof-level. This beam of light was clearly defined, without any fuzzy edges, and of brilliant color. Both Mr. and Mrs. Knibbs had great difficulty describing the exact color, and both said it was like the flame of an acetylene welding torch, with brilliant red, orange and yellow all mixed up together.

"As they watched, the beam slowly disappeared from roof-level to the floor, as if it were descending into the ground, and as it reached the floor, a small dome of light, or ball of fire, formed. This quickly spread, and in less than a minute this section of the factory was ablaze."

Other pieces fitting into the puzzle were a light in the sky in the vicinity of the factory, observed from a point about two miles away, just before the explosion occurred; and reports of a clicking noise preceding the explosion that was similar to sounds associated with other UFO sightings. Putting the whole picture together, Cathie concluded that it was indeed a case of the aerial visitors at work on their power grid and he theorized that the beam of light was a column of pure energy, similar to a laser beam, being implanted in the earth. He speculated that once it had penetrated to the required depth, it might have been reconverted into a material state to form an aerial.

In the summer of 1970, two years after Cathie's book was published, I investigated a case which led me to think along similar lines. It was brought to my attention by Dr. P.M.H. Edwards, a former instructor in languages at the University of Victoria in B.C. who is well known as a student of ufology.

There were two parts to the case. One appeared as an item in the Victoria *Colonist* which said that in the Saanich district, outside the city, police were puzzled about an accident in which a hole was punched through the heavy aluminum roof of a shed on the farm of F.H. Adams, a trainer of horses.

"In addition to the roof," the item said, "the mystery missile split a one inch board and shattered a 4 by 10 beam. Police said the hole in the roof was about six by three inches. 'We were breaking horses most of the day,' said Adams. 'My son was the first to notice it. We went to the shed to get a piece of equipment and he looked up and said, "God, look at that Dad!"

" 'We looked all around but couldn't find anything. The pieces of wood from the beam are lying on the ground, but there is nothing to show what caused it.' "

Dr. Edwards reported the second part of the case (assuming it was related) in a note he sent with the clipping. He said Daryl Brachat, a recent science graduate from the university, was in bed that same night when "suddenly he saw a line of yellowish-white light about the width of a fingernail at arm's length (perhaps three or four miles away?) It quickly darted eastward, stopped dead, then retraced its tracks partway, then went off fast southward towards Port Angeles. Mr. Brachat has often seen satellites that might pass overhead. He knows that what he saw was by no means a satellite."

To see if there was anything more to the first incident, I drove to the Adams' horse farm and was allowed to look around for myself. I found that beneath the hole in the roof the floor of the storage shed, obviously old, was worn or broken away in places, exposing dark soil underneath. While the dirt was fairly loose, it was still firm enough to block the fall of anything dropped from above or at least show a trace if the object had plunged below the surface. But, as Adams and his son had already noted, there was nothing in the shed after the hole was discovered that had not been there before, nor had anything left a mark on the soil.

As a possible answer there remained the chance that a chunk of ice had fallen from the wings of a passing aircraft, broken through the roof and melted on the ground before it was discovered. While the heat of that clear day in mid-July could have destroyed the evidence, that same condition also eliminated the likelihood of ice being formed in the first place. Besides, the only clue I could find suggested that the missile had nothing to do with ice. Instead it was something of such searing heat it had affected the metal of the roof where it entered.

As part of the manufacturing process, the aluminum sheets forming the roof were marked with evenly spaced parallel lines. Normally, anything puncturing the roof would not have affected those lines other than to break them. Yet around the hole in the Adams' shed the lines were wavy and irregular and the nails were loosened as if the metal's very composition had been disturbed, seemingly by exposure to intense heat for an instant longer than it would take a free-falling object to crash through. As I looked at the damage I had a picture of a shaft of light passing through the metal under pressure, and causing stress that snapped the old woodwork, to disappear in the ground below in the manner described in Bruce Cathie's book.

Was the "line of yellowish-white light" seen over Victoria a UFO aerial, carried by a craft invisible in the darkness, that was planted the same night in the Adams' farm? That question, with all its unlikely ramifications, would not get a second glance from an orthodox scientist. But in ufology we have to work a bit harder. Ours is not a laboratory where time is of little concern and the much sought duplication of results in an experiment can be awaited with infinite patience. Instead, for us a quick unexpected glimpse must usually suffice, and from that we have to devise a picture of probabilities. Remarkably, after years of glimpses, an obvious and convincing pattern has emerged. UFOs are real and they have a purpose.

In the case of the Victoria sighting and the damaged shed, a connection seemed quite possible in terms of what we know about UFO behavior, particularly as both incidents were unusual. But as I was thanking Adams that summer afternoon for allowing me to visit his farm, he made a remark that removed some of the guesswork.

"You know," he said, "there was someone else who came around thinking there must have been some kind of spaceship around here but I can't help much when it comes to things like that. Sure, you were welcome to look around, but I just don't know anything about it.

"The only thing is, though, about a year ago my daughter, Shirley, and I were standing outside one evening when we saw a ball of light come over that hill over there." He pointed to a small hill about two miles east. "It was red, about the size of a grapefruit, and it was moving very slowly. I asked Shirley what she thought it was, and she didn't know. I didn't know either, but whatever it was, I'd say it was looking for something."

Perhaps that something was a place to do some work.

With this indication that alien visitors were interested in the area, the case of the damaged shed would take a definite place in the story of aerial planting were it not for one thing — no booming noise was heard. In fact, there was no noise of any kind.

Possibly, of course, the booming is a secondary result of the operation, if aerial planting is indeed being carried on. Even Bruce Cathie, in discussing destruction of the foundry at Auckland, speculated that the noise was caused first by the roof collapsing and next by the reinforced floor cracking.

However, since part of the picture is missing from the Victoria case, it might seem logical to discard the whole. But here again we must remember that in ufology something is better than nothing. In this instance it seems likely that some organized activity was afoot and, rather than forget the whole thing because an item is absent, we must be grateful for what we have and look for a purpose.

In another case I investigated, along with Bill Allan, the noise was there but the damage was missing and no shaft of light was reported, yet the rest of the activity was too much to ignore in a field where pickings are so thin.

The place was the region of Prince George, a thriving center of the forest industry in north-central British Columbia. Because of its rapid growth in recent years, Prince George has been known as a "boom town" and on the night of July 4, 1977, it was literally just that. For reasons that remained unexplained a series of shakes rattled windows in the city that night. The effect was like that of several mini-quakes, yet no earth tremors were recorded on the local seismograph. Although the shakes dwindled away, what may have been a climax occurred when a thud like a smothered explosion was heard by residents in the heart of the city.

So here we are back to a resounding noise heard amid mysterious ground shudders, similar on a small scale to the Atlantic seaboard incidents. Conceivably in this case, as well as in the others, the effects were caused by something plunging through the earth's surface. But what about the agent of delivery? Surely if this was another instance of UFO capsule planting, some sort of unusual aerial activity would have been seen about the same time.

Brian James

And so it was, but it was seen a few hours after these events, not before or simultaneously. However, since the craft involved operated in total silence, even at the point of near collision with its witness, it could have flown over Prince George earlier that night, its lights extinguished, without being noticed by anyone. Moreover its appearance and behavior were so extraordinary, there was no doubt our visitors had special business to perform in that area.

The witness was a young man named Kirk Alore who was employed in a supermarket in the neighboring town of Fort St. James. Kirk was driving home from Prince George in the early morning hours of July 5 when his attention was caught by a brilliant red light in the air about a mile ahead and to his right. He added, significantly, that it was above some hydro-electric towers.

I just kept driving on and looking at it, and then I saw a car coming. It was about half a mile away. Then I noticed this red light move. I kept staring at it and it just sort of shot right over the top of the oncoming car. I just stared at it in amazement. Then it shot towards me, when the oncoming car was about 300 yards away, and it was huge! It took up about two highway widths. At that instant the red light lit up the interior of my cab and I got strange sensations, like my arms fell asleep and my legs fell asleep and they felt like pins and needles. Then I swerved and missed this thing because it was coming right towards me.

It all happened so fast I just don't remember anything after. I sort of blacked out, I guess. According to our watches it was 15 minutes later or so (making it about 4:30 a.m.) and this guy was opening my door and shaking me. He was asking if I was all right. I said, "Yes, I'm OK, I guess." But I was sort of sick at the moment and was aching all over. My arms and everything were sore. Our cars were about two feet apart.

He said, "A strange thing happened to me. I was outside my car when I came to." We could find no skid marks and I was doing at least 70 miles an hour when I swerved into his lane.

We went to the hospital in Vanderhoof, both of us, and they checked us over and there was apparently just shock. That's all it was. I never did get the other guy's name.

Bill Allan later tried to confirm this at the hospital, without success. However, since it was not necessary to admit the two men, even as outpatients, possibly no record was kept.

Asked for a description of the object, Kirk said:

It was oblong; the outer shell looked more or less like alligator skin. It had little protruding wings, except they were rounded. Underneath between the wings it had a circular thing, like an electric razor. You know, the discs? That's what it had underneath, one turning slowly inside the other, which was turning the other way. It had a dome on top with different colored lights inside — green, red, yellow, blue, all like little dots. Then it had just one straight beam of red light when it shone the red light in my truck.

He said he stared at the thing for what seemed like "the longest time" when it was over the top of the other car. "I could see everything in great detail."

Kirk estimated the diameter of the "circular thing" underneath was about 20 feet and, while the explanation was not quite clear, we gathered the inner circle was about 15 feet in diameter. They revolved about one center.

He said the surface of the object reminded him of an alligator skin because of its "texture — the ripples in it — and its color was dull, somewhat like an alligator skin."

The other man told Kirk he had first seen the object in his rearview mirror, and as it moved in front of him he too had prickly sensations. After that he didn't remember a thing except seeing Kirk's headlights and the bottom of the object.

While the details of this experience may seem irrelevant since there was no observed connection between it and the earlier shakes and noise in Prince George, I mention them because first of all they describe the presence locally of a flying object that was quite singular even in the bizarre world of UFOs. The alligator-skin texture of its fuselage, its tiny wings and its great size must have given it the appearance of a vehicle that, for lack of better definition, may be thought of as some kind of space freak on special duty. Second, there is one detail at least that does much to establish the honesty of the only reporting witness. It is his reference to a system of contra-rotating fans on the craft's belly. In my files I have a report on a little-known case far removed from Kirk Alore's in place and time but containing a detail almost identical to that of the "circular thing, like an electric razor" which he described. If he had dreamed up a device so similar, I think it would have been more incredible than seeing it.

Now, if it appears that in this discussion of mini-quakes, sound effects and an aerial monster we have forgotten the opening theme of our visitors' interest in electric power, we must remember that no one progresses in ufology without at least a modest degree of imagination. By using some of it here we can visualize spears of concentrated energy being driven into the ground at widely different points, sometimes with explosive impact, in the manner theorized by Bruce Cathie. Since we know from numerous sighting reports that our visitors can manipulate the energy of light apparently in any way they wish, casting it into beams that bend, break and perform a maze of other tricks, we must assume they can handle electric energy just as easily. With that in mind, a picture of long electric bolts being pressed underground falls readily into place, and so do the attendant results.

Finally, still using Wido Hoville's report from Quebec for our line of study in these two chapters, we now take a quick look at whatever intelligence, robot or otherwise, lies behind the machinery of this alien affair with our electric power. It has to be quick because our usually secretive visitors are still more secretive when engaged in a business so obviously vital to their presence here.

One who had a particularly good glimpse of their presence at such a time was the night supervisor of a radio-TV-link station in Angermanland, a province in northeast Sweden. The date was March 12, 1969, but for fear of ridicule, the witness, Koell Naslund, said nothing about his experience until interviewed more than three years later by *UFO-Sverige* which carried the account in a special report dated October-December 1973.

The witness, described as a respected man without any previous knowledge of UFOs, had just come on duty at 6 p.m. when the station alarm sounded signifying a possible overload in the transmitter. Seeing nothing wrong, he reset the alarm only to have the same thing happen three or four more times. Going outside, he found the weather clear and cold, with northern lights, but nothing unusual. He then checked with the main station and was told its alarm had also acted up but not to the same extent.

UFO-Sverige, providing its own English translation, — continued:

> Mr. Naslund said, "I don't know for sure how long these alarms went on but around 6:30 it was like someone told me to 'Go out!' It was some sort of feeling."
>
> He went to the door, opened it, looked out and stared! He saw the "thing" he will never forget in his life. About 15 meters (50 feet) from the station, in a bare dip in the ground with only some lonely trees and vegetation, was a gigantic cigar-shaped object. Mr. Naslund estimated the length to be 150 meters (500 feet). The object filled the dip in the ground below the station house. The front part of the object was at the same level as the ground outside the station. In this part of the object Mr. Naslund could see a tunnel inside. In the tunnel was a bright gray-blue light while the rest of the object shone with a dim gray light. Close to the object and inside the tunnel Mr. Naslund could see about 10 creatures which he described like this:
>
> "They resembled boxes someone had sprayed with something. They also resembled clouds, and they floated above the ground. I would say they were about 130 - 135 centimeters (four feet plus) tall and 30 - 40 centimeters (one foot plus) wide."
>
> According to Mr. Naslund, the tunnel in the object resembled a three-dimensional movie screen. "There was depth but I couldn't see right through it." In the tunnel opening three or four creatures were floating around. The first thing that occurred to Mr. Naslund was that he had to get to the phone and call the police about what was happening. He was about to return inside when he found he couldn't walk! He said:
>
> "I got this eerie feeling again, that there was a job for me. . .I should keep the door open for them. I couldn't move an inch from where I was. I could only watch. There were seven or eight creatures who floated into the link station, and they passed me by only a few decimeters! I tried to touch them but I couldn't. Some force held me back. I got some strange feeling that there were creatures in the square-shaped wraps they had and that they wanted to protect themselves from a closer look. At the same time a feeling came over me that I should be calm since they were not harmful."
>
> "Was that a feeling or did you hear the voice in your head?"
>
> "I can't explain it any other way than a feeling. It was no voice inside my head."
>
> The box-shaped creatures didn't move very fast. They floated ahead at the same speed as a human walking. They didn't move otherwise and they had neither arms nor legs. Mr. Naslund did not hear any sound when they moved nor did he smell anything.
>
> He estimated the craft's height at five to six meters (16-20

feet). He could not see the lower part since it was covered by the dip in the ground. Its contours were as dim as the creatures'. He said:

"There was light from the object, and the light trembled as if it was shining through warm air meeting cold air. The object was buzzing like a transformer station, and that was the only sound I could hear. . .I was so influenced by the creatures I had no will of my own. I felt as if I was directed by them. I think they were inside the station for 5 or 10 minutes. I'm not sure but I don't think it was longer than that."

Mr. Naslund said he was fully conscious during the whole incident but was unable to do anything of his own free will. Also he said his watch continued running normally.

Before the object took off, the creatures were "sucked" into the tunnel which then closed after them. In a moment the object rose and flew off in the direction of Vannes City where a transmission mast was visible nine miles away.

As the object left, the witness phoned the police in a nearby village who seemed to take his report seriously but could not see anything strange in the sky, though perhaps by that time the craft was too far away. Mr. Naslund said nothing in the link station had been touched. He continued:

"I went back to the station next day to see if there were any traces from the object, but the ground was hard with frozen snow. The only clue was a pine tree which had its top cut off where it was about three inches thick. It was cut off at the same level as the ground outside the station."

It appeared from Mr. Naslund's answers that the craft did not lie on the ground in the dip itself, which was about 15 meters deep, but hung in a lowered position parallel to the ground outside the station.

At nine o'clock on the same evening as the incident a man driving in the vicinity told a local newspaper he saw a white, intensely bright object in front of his car. But *UFO-Sverige* could not find the man to verify this story.

At the end of this account *UFO-Sverige* said that later a news clipping dated January 1969 (two months before the incident) came to light reporting the experience of someone in the same province who saw a creature remarkably similar, though larger, to those that entered the link station. The witness was Gottfried Olsson, 62, who was looking out the window of his home in Domback village one night when he saw a "lighted object, big as a human" standing beside the highway.

"The object was not an ordinary human but still resembled one," he said. "It was about the length and width of a human, while the shoulder part was clearly marked and so were the lower and squared lines, somewhat like a box! Unlike its body, its head had somewhat dim contours and was 'muddy.' The object stood perfectly still while I watched it for three or four minutes, and it radiated a strong red-white light except for a vertical, blue-colored middle line. Suddenly there was a brilliant flash and the object disappeared. I could not say where it went — it was gone just like that!"

It will be noted that this creature had a resemblance not only to those seen at the link station but also to those seen five years later in Quebec during the incident reported by Wido Hoville. There the head sections also seemed to leave little or no impression and, as well, the body portions radiated a strong light. With matching details of this sort in cases emanating from such widely separate sources, we are obliged again to accept the probability that here is something stranger than fiction.

However, despite the extraordinary nature of these two incidents, and the illuminated figures' apparent lack of concern about being seen, it was noticeable that an important secret was being kept from the witnesses: namely, the identity of the mind behind all the action. This was most evident in the link station when Koell Naslund felt an influence forcing its commands on him. This conquest of instinct is a favourite means of communication for our visitors and, silencing any back-talk as it does, helps explain how they retain their anonymity.

Another trick is simply to make the witness forget where he is or what he is doing. We see this used in the second of the two cases mentioned after Wido Hoville's report (in the preceding chapter) as rare examples of UFO occupants seen right on the site of a power installation. This time the occupants had a human form, at least to the extent they were visible.

The scene was near Radium, British Columbia, beside the Columbia River in what I think of as the "Playground of Gods" because of all the sightings reported to me on visits there. The action started one night in July, 1975, as Frank Slotta was driving his wife, Lynn, home from work at Radium Hot Springs Lodge. Also in the car were their daughter and two young men who lived nearby. All five saw what happened when they were a mile or so from the lodge.

The Slottas described the incident to Lillian and me while we were on a visit to the area three months later.

"That light's moving!" Lynn had said, pointing to what at first appeared to be a star above a ridge to their right. Frank thought it was just because the car was moving, but he stopped where they could have a better look and they rolled down the windows.

"The light moved a little more," Lynn told us, "blanked out, then all of a sudden it was over the hill above the power station." (She was referring to a transformer unit beside the highway.)

It was a curious incident but not quite enough to convince the group, anxious to get home, that they should stay there and watch. After all, lights at night are usually open to various normal explanations. As it happened, however, Frank had to return along the same route a few minutes later for his job at the lodge where he was

employed as night watchman. So he was on the lookout for anything strange. . .and as he came around a bend in the highway near the power unit, there it was.

"I looked up and saw the light real close, above the slope behind the power station," he said. "So I slowed right down and then I saw there were three strange-looking sort of orangey beams shining down."

By this time Frank had come to a stop. Then realizing he was parked in the middle of the road, he pulled to one side to continue watching.

As he stared, wondering momentarily why other cars did not stop, the source of light came down until it was among the trees and he could discern its shape.

"It was a double-shaped deal about 40 feet wide, maybe more. On top there was a part with rectangular windows, lit from the inside, and below there was a larger part with lights shining out from the rim."

In a drawing he made while we talked, Frank also showed three lights shining down, and it was evidently these that provided most of the illumination. He was surprised that he "could see everything, even the needles on the pine trees."

But it was the top part of the vehicle that mostly held his attention. About half as wide as the bottom part, it was six to eight feet high, with a long window in the center that seemed to hypnotize him.

"For some reason I couldn't take my eyes off it," he said. "While I looked I saw a shadow moving by as if there was a person inside, and it seemed from the way it went back and forth that maybe there was more than just one."

Although no details were visible, the shadows impressed Frank as being like those of humans in shape and manner of movement, and with his window open he knew he was not deceived by any reflections from the car's interior. Also he noticed that in the still night air the strange craft made a rhythmic swishing noise. He imitated this for us through his lips, and it had an uncanny resemblance to the sound reported by witnesses in other cases.

At last, realizing he would be late for work, he looked at his watch and saw that unbelievably he had been there for more than an hour.

"I thought it was more like ten minutes," he told us, obviously still puzzled.

But continuing on to the lodge about a mile away, he allowed himself one more look as he left the car. He knew it was unlikely he would see anything as the object had been on the other side of the ridge near the highway. Miraculously, however, there it was in the distance, looking even more brightly lit than before. Apparently it had moved upward as he left, almost as if it wanted to exchange one last glance.

Frank phoned home as soon as he entered the lodge, and his wife and daughter remembered the time was about 1:30 a.m. He had intended to start work at midnight!

Thinking the rest of the night about what he had seen, Frank returned to his place by the highway at the first opportunity next morning. As he had expected, there was no sign of the craft but curious to see if it had left any traces on the ridge, he found a back road that took him part way up to the top and laboriously he climbed the rest of the way until he was at the exact place of his sighting.

"But there was nothing," he said. "I spent an hour looking around but there wasn't a sign it had been there. So it must have been in the air all the time even though it looked as if it was on the ground."

However, Frank did make one discovery of interest. He found there was a power-line pole at the spot where he figured the object would have been. It appeared, in fact, that the craft must have made its hovering stop right next to the pole. With the power transformer a short distance away at the foot of the ridge, this meant our visitors had once again parked near a good supply of electricity, presumably to help themselves to some of it.

But in this case something was added since the power-thirsty tourists exposed themselves as they did at the Swedish link station. Again, however, they were secretive despite their apparent boldness. While the witness blanked out, did they leave their craft to make a connection with the power line, or did something else happen that would have disclosed their purpose beyond any doubt?

While this may seem to be imagining too much, even for a scene that was already unusual enough, we were left with a strong feeling that something still unexpressed after three months had happened to Frank that night. It seemed to stem from what he saw behind that long center window which, as he told us several times, had fascinated him.

His wife, Lynn, believed he had told us everything he could remember and we were sure she was right. Frank, a very pleasant man, had been open and cooperative. Yet something about his experience, perhaps beyond conscious memory, appeared to defy expression.

This thought also occurred to Bill and Yvonne Allan who joined us on our visit to Radium. As we had told them ahead of time about Frank's encounter, Bill brought along a selection of his UFO slides which he showed one day when we stopped in at the Slottas' place where they ran a small farm.

The effect of the pictures on Frank was obvious at once. He stared at them as if lost momentarily to everything else around him, a reaction that encouraged Bill to mention that hypnosis sometimes helped UFO witnesses to overcome memory block. Would Frank care to try it?

Frank did not turn the idea down but arrangements out there on their remote farm would have been difficult and the idea was dropped. So we can only suppose that the observed occupants of the mysterious craft did indeed make contact with a power-line that night. If they did not, why did they linger for more than an hour just inches away from the source of supply while their witness looked on, mesmerised?

Chapter Seven: On The "Meteor" Trail

Lillian Magor walks along the shore of the *Playground of the Gods*, an area in which frequent UFO sightings have been reported.

In February, 1974, news services were alerted by a story in the British scientific periodical *Nature* (vol. 247) reporting on the near-miss of a potentially devastating "meteor" over western U.S. and Canada, August 10, 1972.

Unaccountably held as classified information until its release 18 months after the incident, the report by a team of five scientists summarized the event as a rare occurrence that would probably happen only once in 10,000 years. With an estimated weight of 1,000 tons and a diameter of 13 feet, the object was thought to have the destructive capacity of "the nuclear weapons which destroyed Hiroshima and Nagasaki."

The press report added:

"Refined measurements from satellite tracking of the meteor show that at 2:30 p.m. local time it sped along a course from Salt Lake City, Utah, north to Calgary, Alberta, reaching a low point of only 26 miles above Eastern Idaho, just south of the Beaverhead Mountain range.

The Rocky Mountain Trench or the *Playground of the Gods* looking north.

B.C. Government Air Photograph

"'If it had been at a slightly lower altitude, the damage would have been very extensive', the scientists wrote. 'The meteor very narrowly missed hitting the earth. Fortunately meteors of this size are exceedingly rare.'

"The meteor became hot enough over Utah for detection by infrared instruments aboard a satellite and cooled below the detection threshold over Alberta. It zoomed out of earth's orbit again apparently without breaking up."

It seems at this point, several years after the event, that the "meteor" will remain secure in scientific records as just what it was said to be, a meteor. Even before the article in *Nature* put a stamp of finality on the case there seemed to be remarkably little doubt, even in general discussion, that it was anything but a natural phenomenon, albeit an odd one.

However, as a frequent visitor to the "Playground of Gods" where the meteor — or something else just as extraordinary — was seen in flight that same afternoon of August 10, 1972, I have strong reasons to think otherwise.

A few miles away in the "Playground", where the Columbia River forms part of the Rocky Mountain Trench, Frank Slotta had his encounter with an unusual hovering craft that contained one or two shadowy crew members, and there have been other sightings just as strange. However, the one that gave the Playground its name in my mind occurred on a clear spring morning in 1963 when a giant glowing ball was seen bouncing above the Rocky Mountains of the Trench's eastern wall, almost hitting the peaks. Witnesses told me how they watched spellbound as the brilliant object flew south toward Montana in its soaring dance. As usual in that place, the "gods" were having a ball.

But spectacular though it was, their game on that occasion may also have been brief because it never made news in the manner of the so-called meteor nine years later. Nor was the bounding ball seen by so many in so many different ways, and it is those variations in the later act that concern us here. How could a meteor perform all the tricks attributed to it by various eye-witness accounts?

One spectator with a front-row seat on "meteor day" was Mrs. T.N. Weir, editor of the weekly Lake Windermere *Valley Echo* in the Rocky Mountain Trench. Mrs. Weir wrote about her experience months before the *Nature* article appeared, which she was unlikely to see anyway.

She said at the time it happened she was sitting on her house patio in Invermere, beside Lake Windermere, with her brother and a friend. Suddenly an object hurtled into view, heading north.

"The patio looks out toward the apple-trees in our yard," she said, "and the object was so close it seemed to skim just over the tops of the apple trees. It was a beautiful sight. Some who saw it thought it was chrome-coloured, but to us it looked like an irridescent blue, although that may have been due to reflection. In any case, we definitely thought it had a metallic appearance."

When I saw her on a visit and asked for more details, she described the object as about the size and shape of a small jet but there were no wings, port-holes or markings, and there was something else missing — even though it was barely a stone's throw away, there was no sound! Within seconds after it passed, the newspaper phone started ringing, confirming that something strange overhead had been widely seen throughout the district.

All that was remarkable enough, but the gods in their playground had another trick to perform. An hour or so later there was another flurry of calls to the newspaper to report what was apparently a second object similar to the first. It too was said to be flying low and headed in the same direction.

A few years later on yet another visit to the area I learned from Mrs. Belle Ede, who succeeded Mrs. Weir as editor of the *Valley Echo*, that she, her small son and a friend of his were on the Lake Windermere beach that eventful afternoon when a flying object with a flaming tail appeared overhead. At first she thought it was a jet plane on fire, but it kept going and disappeared behind the mountains.

"It was going in the direction of Calgary," she said, which meant it was flying east, not north as in the case of other sightings. Since those mountains run north and south, her estimate of the object's direction must have been correct otherwise it would not have crossed them.

So just how many "meteors" were in the air that afternoon?

In the first place, the object reported in *Nature* should not have been seen over Invermere at all. Calgary, said to be at the end of its visible flight, is 100 miles almost due east on the other side of the lofty Rockies. To be seen simultaneously from Invermere and Calgary, the object would have to be at a height somewhat greater than the trees of Mrs. Weir's apple orchard!

Secondly, there were reports that same afternoon of a "meteor" seen far east of Calgary. Residents of Gull Lake, Saskatchewan, 200 miles away, said they saw a flying object "shaped like a basketball bright and shiny in colour as if of silver metal." They said it disappeared in mid-air.

Finally, in the very area of Alberta over which at that instant the meteor was supposed to be climbing back into space, there were witnesses who saw one or more flying objects behaving in a manner quite unlike that described in *Nature*.

Two witnesses were Barrie Jones and his son, of Calgary, who were on a picnic that afternoon when they saw something shiny streaking in their direction from the south. When Lillian and I called on Jones in his Calgary office to talk about the incident, he said he guessed the object's height was 4-5000 feet. (The scientists' estimate for the height of the meteor at that point was more than 36 miles.) A camera fancier, he was wearing a Swiss-made 35 mm wrist-camera, and with this tiny device he obtained a shot of the object as it approached, another as it came alongside and a third from the rear. The first picture was so good that it was reproduced in newspapers. Obviously since he obtained clear pictures with an extremely small camera, Jones' estimate of the "meteor's" height was more accurate than the scientists'. Or was it the same object?

In reporting his experience to the Calgary *Herald*, Jones made a point that added to the mystery.

"It was very shiny", he said, "with four jets that were multi-colored."

A shiny meteor with four jets? Yet Jones timed his sighting at 2:29 p.m., making it virtually certain he saw the same object that was later discussed in *Nature*.

However, other witnesses in that area just as firmly put the time at 3:00 p.m. or slightly later. One was Mrs. Weir of Invermere and another was D.S. Douglas of Granum in southern Alberta, who claimed there were at least two strange objects in the sky that afternoon because he saw them together! In a letter to us he said:

"I still remember that day distinctly. It was August 10, 1972. The weather was hot and dry, 80 to 85 degrees, and the sky was clear.

"It was about 3 p.m. when I sighted an object coming over the horizon due south, leaving behind it a trail of smoke. I ran to my camper which was only a short distance away, jumped in the back and got my binoculars. As I stepped out the object was directly in line with me, due west. At this point I checked the second hand on my watch.

"The object was about 70 degrees above the west horizon about this time. On looking through the 7 by 50 power binoculars, I saw the object was burning brightly like a welder's acetylene torch. The burning was taking place at the rear of the object, not the front. It appeared to be cigar-shaped and quite large, leaving behind it a trail of smoke.

"I continued to watch it until it was about 20 degrees above the northern horizon. At this point the burning stopped and all I could see was a little speck, which then disappeared. A trail of smoke was visible from the southern horizon almost to the northern horizon. It appeared to have dips in it as if the object was in fact adjusting its own altitude.

"Approximately 90 seconds later a sonic boom was heard which sounded like rolls of thunder. I then continued to look at the smoke trail. At this point I observed a saucer-shaped object slowly moving along the smoke trail. I watched this saucer for about 10 seconds, then it disappeared in a flash."

If the "meteor" was not a meteor — and witnesses like these made it sound questionable — what was it? Perhaps without realizing it, millions of television viewers were given the answer in December, 1974, when NBC carried a documentary on the UFO question. A highlight of the show was a remarkable filming of this same object taken by an observer who was quick with his camera. Although it was described as a natural phenomenon in accordance with the accepted scientific opinion, viewers must have found it hard to see it in that light. What they saw was not a wild flaming thing that would be acceptable as some kind of fireball. Instead they were looking at a gracefully shaped object with a pointed front section and tapered body followed by what seemed to be a long, evenly slender stream of exhaust. It could easily been anyone's conception of an alien spaceship, with a special thought for that "smoke trail."

Normally UFOs do not leave a trail of any kind. It is as if they operate in a self-created air pocket where they cause no outside disturbance and in which even sound is contained.

We can speculate, then, that since the object of that eventful August day in 1972 did leave a trail — and a conspicuous one at that, stretching from horizon to horizon and, by some estimates, lingering for several minutes — it was for a purpose. Moreover a clue about that purpose may lie in the presence of the second object which eyewitness Douglas, for one, said was following in the smoke trail. As the second object was unmistakably the more familiar flying-saucer type, the whole action seems to have been related to general UFO operations. This in turn poses a few questions that could touch on the answer of what it was all about.

Was the second object literally on the scent of the first? Was the "smoke" deliberately made to linger so the second object could indeed smell and analyze it? Was this some sort of pollution test by an extraterrestrial intelligence interested in what we are doing to our atmosphere?

The scene of action may also have had significance. Regardless of how many objects were in the sky that afternoon, the majority of sightings lay along a swathe stretching north from Utah through Idaho and Montana and into British Columbia and Alberta, all of which are close to Rocky Mountain country or within it at places. As we have seen in these pages this appears to be country favored by the aliens, and evidently they know it well. What better place to carry out a long-ranging test?

If auxiliary craft were needed, which could account for the variety of sightings, these could be provided in the mountain thoroughfares with a minimum of public attention. (One report which I was unable to check to its source said that on the same afternoon an elongated object was seen to fly step-by-step up the side of the mountain near Windermere in the Trench. When it reached the top it vanished.)

Pinning a possible air-test by space visitors on this incident is not just wild guesswork. Any close study of UFO occupants suggests they have to cope with the problem of breathing our atmosphere. Since reports of aliens without a vehicle in sight are comparatively few, it appears the visitors generally prefer to remain inside or at least near their craft. Yet the assortment of their headgear, ranging from thick masks to cloth-like coverings, indicates they manage to handle the problem in various ways, with some being so advanced that no mask at all is needed. Possibly they have chemicals to adapt.

Even so, the problem is obviously a continuing one and writers often bring it up. One noteworthy reference is attributed to the late Gen. George C. Marshall who is said to have remarked in 1955 that authorities in the U.S. "have established the fact that flying saucers are manned by visitors from outer space, and these visitors are trying to work out a method of breathing and staying alive in our atmosphere before landing and establishing contact."

Whether or not the general ever said that is anyone's guess at this late date, and in my files no source is given for the quote. But true or not, the words attributed to General Marshall took on remarkable significance soon after the flight of that "meteor", for in the early summer of 1973, approximately 10 months later, UFOs began to appear over North America on a scale without precedent up to that date.

As noted in chapter one, that year marked the fourth time our visitors had organized a "flap" after the series opened in 1947 and, as before, their performance coincided with a new crisis in the Middle East. This time it was the outbreak of the fourth Arab-Israeli war.

However, as the 1973 invasion developed it became apparent that a new twist was added. Almost literally overnight the aliens began to make landings and, to judge from tracks and grisly traces they left behind, they did so with a boldness they had never shown before. Also, again in a way that was different, they were concentrated overwhelmingly on a particular area, which was the cattle country of midwestern North America.

In a few years this bizarre new phase of UFO operations became known as animal mutilation. That name alone seems to suggest that for some reason the visitors have made a drastic departure from their enigmatic but generally peaceful approach to the affairs of this planet. However, as the assaults continue sporadically into the 1980's, still without a human connection being established, their purpose remains a mystery. Until that is changed, we cannot make judgment against the aliens, even assuming they are responsible. Awful though it is, their cruelty to animals which usually involves the extraction of living parts, leaving the victim to die, may be for a purpose which in their minds will benefit mankind. Our own history of medicine is full of such practices.

We shall examine this subject in the next chapters, keeping in mind that one major point already seems evident. Whatever our visitors' plans may be for this operation, they were made with notable care, if we can agree that the so-called meteor of 1972 was nothing of the sort.

The path of this object cut well into the cattle lands that we will discuss. After a period that presumably was enough for aliens to make a decision in space, farmers in that broad area began to notice strange black craft hovering overhead which, perhaps for lack of a better description, they called helicopters. At the same time some of their livestock started unaccountably to disappear. The aliens' (?) assault on cattle had begun.

We open our study of that assault with a look at the first mutilation case reported to me in the Playground of Gods, which happened within 20 miles of the path of the "meteor." But first a word about more recent conditions there may be of interest as Playground activity seems to typify what is happening elsewhere.

When Lillian and I made a stopover there in the fall of 1982, and spoke to those who had given us so much news in the past, it soon became obvious that conditions had changed. Instead of reports about strange vehicles overhead there was talk about the animal-mutilation problem. It was as if the aliens — assuming they were to blame — had stopped their play overhead to concentrate on grim business below. The talk was scanty, to be sure, particularly where police work was concerned. But there was just enough to raise some real questions.

My latest information came from the ever-helpful local editor, Mrs. Belle Ede, who said that in July, 1981, her paper had a report that a mutilated cow was found on an old airport site down the valley. In a manner typical of such cases a hole had been bored through one ear, and the cow's genitals removed. A police spokesman said that samples from the affected parts were being analysed and promised that the *Echo* would be informed of results.

More than a year later Belle had heard nothing more.

Beings from Beyond

Among numerous top-level quotes collected by Colman VonKeviczky of New York to show the international concern given the UFO question is one by Gen. Douglas MacArthur: "The nations and the world will have to unite, for the next war will be an interplanetary war. The nations of the Earth must some day make a common front against attack by people from other planets" (*N.Y. Times*, Oct. 9, 1955)

That quote is chosen because it typifies VonKeviczky's own concern with alien activities which has become virtually his life's mission. Although his concern was so zealous it led to his discharge from the UN Secretariat in 1966, the restless former major in the Royal Hungarian General Staff did not falter. Soon after his dismissal he founded his own intercontinental UFO network (ICUFON) and he put together what may be the world's finest private collection of UFO photos while compiling data.

As a CUFOR contributor and a good friend for many years, he has helped make it possible for me to present this rare selection of UFO shots taken around the world. Except for a few cases where CUFOR or other photos are used to make a comparison, all pictures in this section are from the ICUFON Archives.

This photo of a disc-shaped craft was taken in March, 1977, as it sped over a marsh near Hamburg, West Germany. (Walter Schilling)

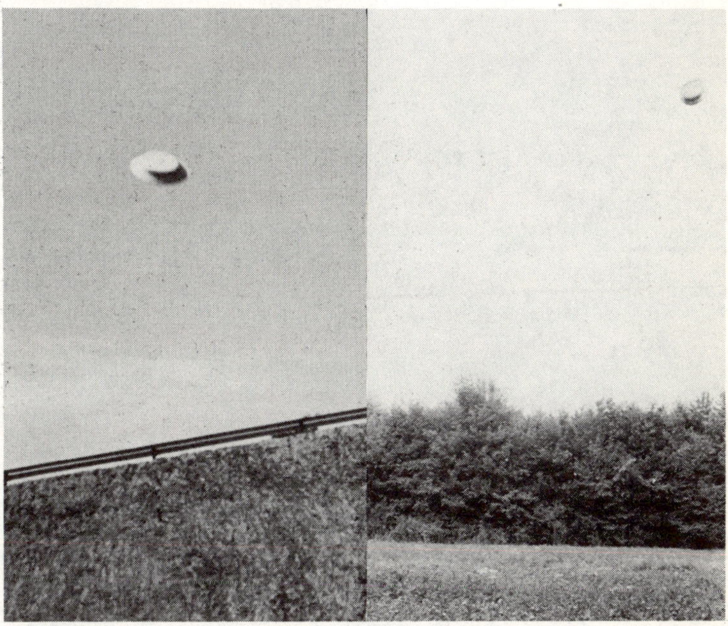

Shipyard worker Jorma Viita photographed this flat-topped circular object (left) as it zoomed above the highway in Yugoslavia in 1974. The photo is part of ICUFON collection. Six years earlier, technician Emile Barnue took several shots of an almost identical object (right) as it gyrated over the woods near Cluj, Romania. (Received from Florin Gheorghita)

In a remarkable series of twelve shots, one of which is shown here, computer programmer Rudolph Nagora recorded the zig-zag flight of a UFO while vacationing with his wife at St. Lorenzen, Austria, in May, 1972. From these photos, Dipl. Engr. Adolph Geigenthaler sketched the path of the object as shown below.

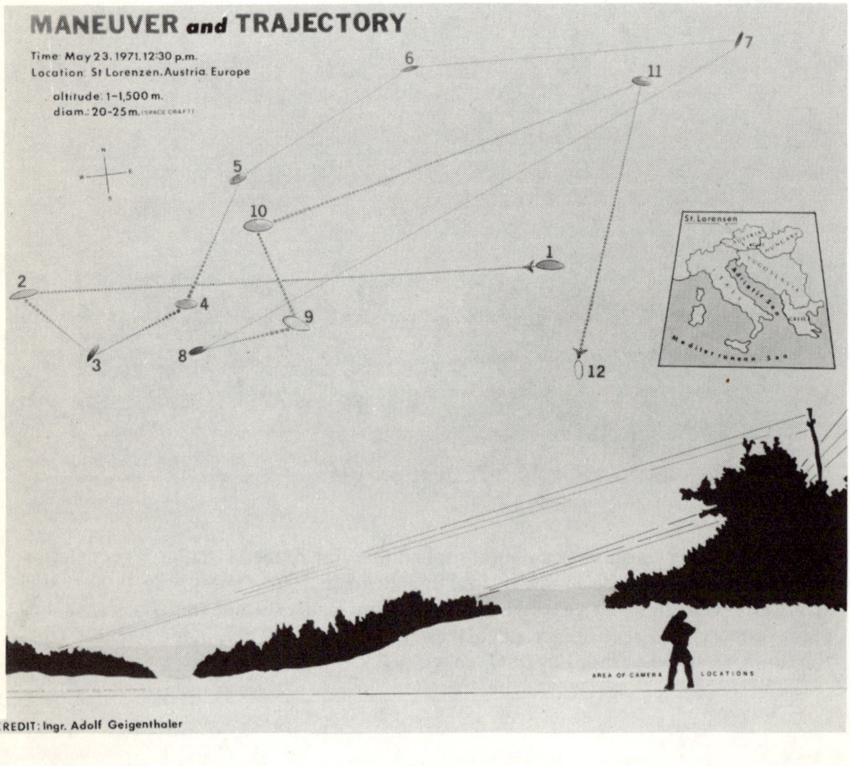

CREDIT: Ingr. Adolf Geigenthaler

Jyri Lina, former Estonian radio journalist, took this photo of two UFOs over Moscow, Russia, one evening in April, 1978. The streaks seem to be the result of time exposure. Due to his UFO research, Lina was forced to find asylum in Sweden.

These six elongated objects, glowing with orange light, were seen over the hills near Lyon, France, in January, 1974.

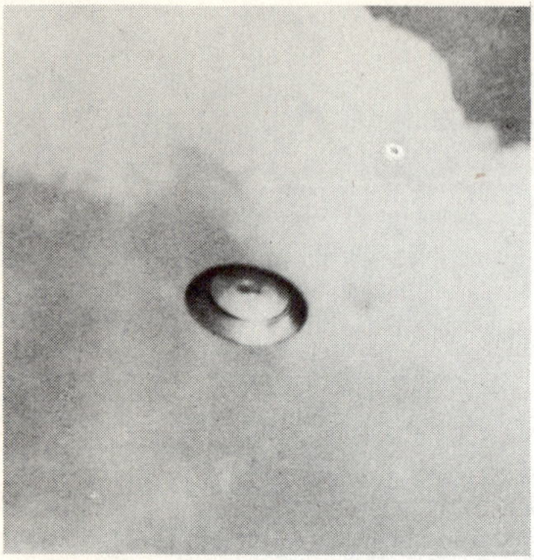

Meteorologist Elisabeth Klarer took this picture of a shiny object while she flew over Natal, South Africa, in July, 1957.

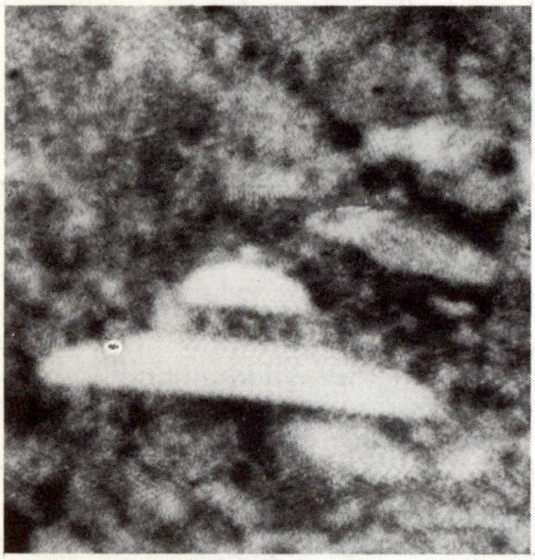

This picture was taken as the silvery hat-shaped craft (right) flew along the Rimac River Valley near Lima, Peru, in October, 1973. (Hugo Luyo Vega)

As described in *Our UFO Visitors*, the object in the lower-left corner of the photo was taken by pilot Alan Laing in 1959 under the guidance of psychic passenger Mrs. Rae. Flying near Banff, Alberta, Laing did not see this object until the film was developed. Except for its color and its squared top, the craft is almost identical to the one above. (Alan Laing)

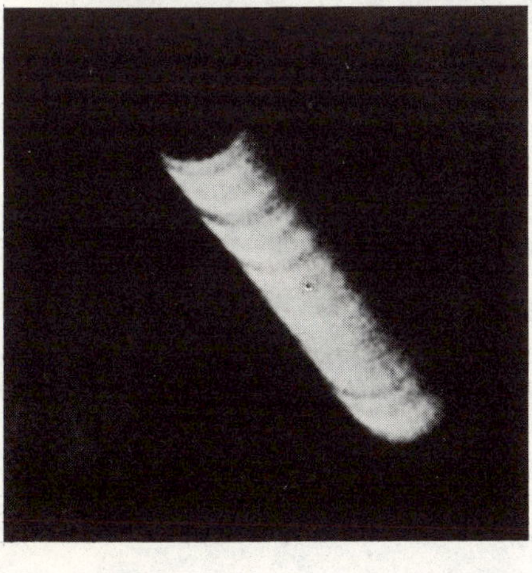

Taken over New York City March 20, 1950, this elongated object was first said to be a time exposure of the moon, but the photo enlargement (left) revealed it to be a tubular-shaped craft. A similar object was seen over Sicily in 1978. (U.S. Air Force)

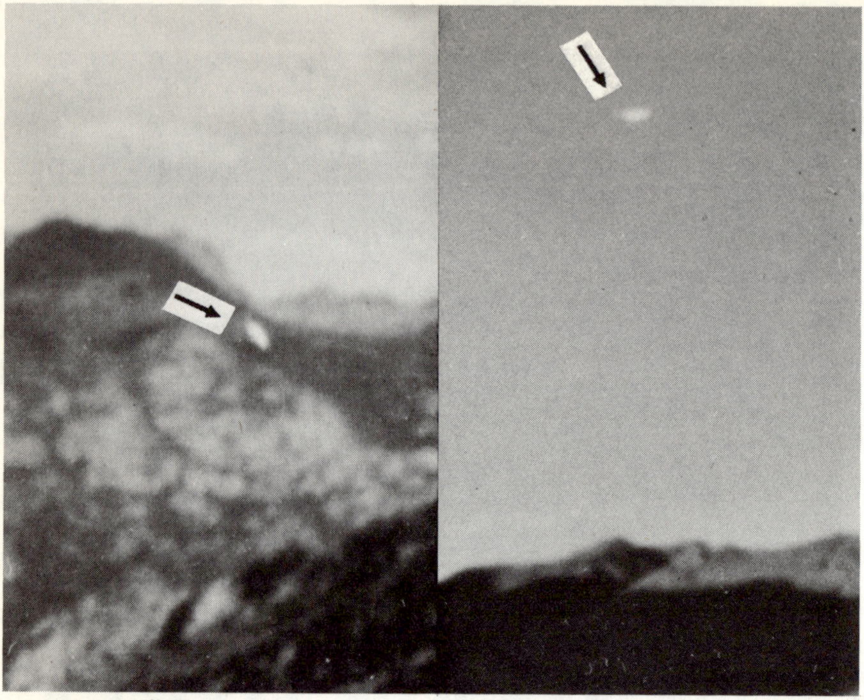

Maneuvering above the chain of hills in Alaska (left), this brilliant white-topped craft prepares to dive into the valley (right). (July, 1978. SBI, N.Y.)

In April, 1970, eight years before the Alaskan incident and about 800 miles away, high-school student Michael Ursulak saw this similar craft flying along the valley near the Peace River in northern Alberta. (Edmonton UFO Society).

In this outstanding photo, the high-crowned object is shown above Hopeh in Shanshi province, China. High-school student Koich Kiryu discovered the picture while organizing his deceased father's photo album. The date of this photo thought to be 1941/42 during maneuvers of the Chinese Liberation Army. On the other side of world in 1975, a similar object appeared (left) at Nanaimo on Vancouver Island. (Lance Willet)

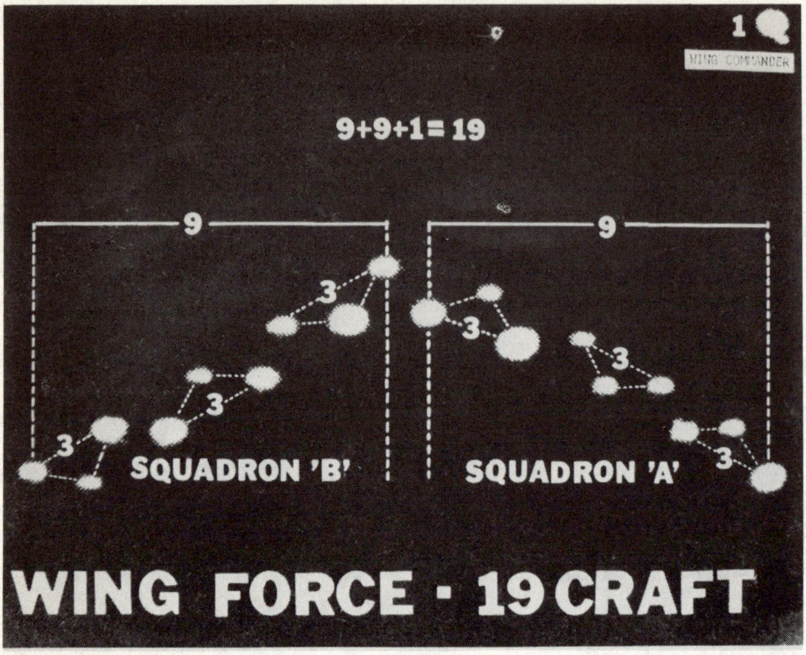

VonKeviczky used his military experience to analyze the Lubbock, Texas light formation seen the evening of Aug. 31, 1951. He saw the group as a *wing* containing two squadrons, each with three flights containing three craft. The wing commander appears upper right. Rejecting the theory that the group was a flight of birds, VonKeviczky said multiples of three are basic UFO formations. (Carl Hart Jr.)

This shot is of a flat disc hovering over a bridge connecting Maji and Shimonosheki in Japan. (April, 1974. Hitoshi Tanaka)

To wind up our global picture tour, we have these remarkable shots of two look-alike crafts seen near the Pacific Ocean. Photo at top was taken from a U.S. military aircraft over the China Sea. (Reprinted from *Overlords, Olympians and the UFO* by Dr. W.G. Allen). In the second photo, a similar strange craft flies unnoticed above the heads of hula girls in Hawaii. (Tsutomo Nakayama, from ICUFON Archives)

Chapter Eight: The Great Cattle Hunt

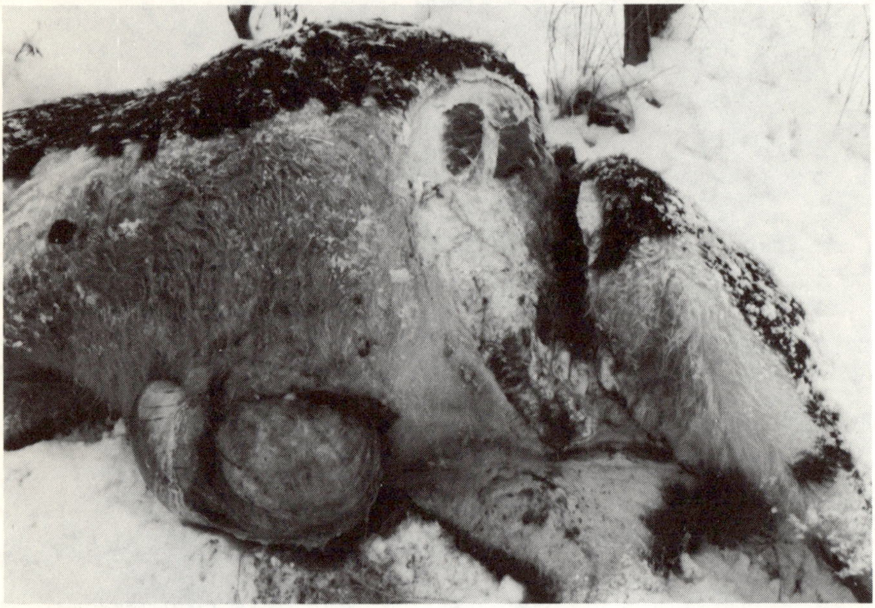

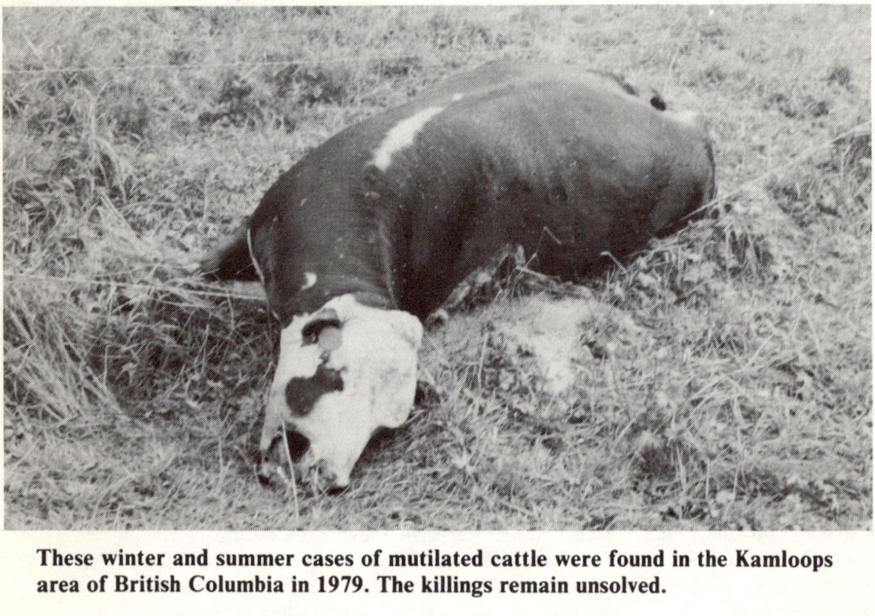

These winter and summer cases of mutilated cattle were found in the Kamloops area of British Columbia in 1979. The killings remain unsolved.

"They say an arrow killed my bull but I don't believe it," Réné Bosc told me. "Whatever it was, it left a large hole in the back of the bull's neck, much too large for an arrow. I think someone stuck a hypodermic needle into the animal and injected something that killed it. Then the hole was made to remove the needle."

We were on a walking tour of Bosc's small ranch near Spillimecheen in the Playground of Gods. At the edge of the field where Bosc had pastured his bull was a brown turgid stream, about the width of a country road, flowing north. More than 100 miles farther along, maturing in plentiful mountain water, it would reverse direction and eventually, far to the south, enter the Pacific as the Columbia River.

The incident Bosc was talking about occurred on a November night in 1978, a few months before my arrival for another look at the Playground. It also involved castration of the animal and removal of half his penis, the sort of work typical of mysterious mutilators, whose field operations, in the most literal sense, extended south through New Mexico and east through the interior plains.

"I think it was done by one of those cults they talk about," Bosc remarked.

As the loss of Bosc's bull marked the beginning of the problem in British Columbia, it received little real attention beyond some questions by police, although to Bosc himself the death of the expensive and newly purchased 1,700-pound bull was a serious setback. While there seemed little doubt he was right in discounting any bow-and-arrow mischief, particularly as it would take a remarkably strong and agile archer to plunge an arrow inches deep into the back of such a large animal's neck, his own idea about a cultist with a hypodermic needle raised almost the same questions.

For one thing, the dead bull was found hundreds of yards from where Bosc usually kept him out of sight from passing cars. For a stranger to locate him at night and then inject something that caused the bull to stumble in his death throes far across the field where the rest of the deed was performed, as recreation of the scene might suggest, would require an almost incredible degree of luck, skill and nerve. Also the carcass was discovered near the main highway. While there was probably little traffic at the time Bosc estimated the deed was done, the grisly business could not have been finished quickly by any human and would certainly have required at least a strong flashlight for help. Yet no passerby or neighbour reported anything strange, as far as I knew.

Finally, although the point was not quite clear, it seemed the animal was drained of blood, which is another mutilation trick, leaving only a small patch on the ground where the animal lay. If that was the case, any human culprit would surely have been still at work in daylight to complete his act. Yet Bosc himself returned late that night to his trailer near the scene and saw nothing.

Although in ufology we may be inclined to give too much credit to our visitors' magic, it was hard to see how all this could have been done with anything less.

While there was obviously much less action in a second mutilation case I heard about, the circumstances were just as mysterious.

This one occurred on the farm of Nick Chomiak at Harrogate, B.C., about 15 miles north of Bosc's place. Having been told of it just shortly before leaving for home I spoke to Mrs. Chomiak on the phone and found her pleasantly helpful in those few minutes.

"We don't know to this day what happened," she said, referring to the death of one of their cattle one night in September, 1979. "The cow must have dropped dead in her tracks because two of her legs were folded beneath her. A sick animal will lie on its side, and anyway this one hadn't been sick."

For the mystified Chomiaks the only clues were as strange as the death itself. An eye was missing, although the lid lay closed over the socket; half an ear had been cut off; and beneath one horn which seemed to have been knocked askew, there was a mark as if something had penetrated the skin, although there was little blood around the wound and no other kind of discharge.

When the RCMP were called in, it was this injury that provided the real puzzle. Unable to determine anything from an external examination, the police severed away the head and, skinning it later, found that whatever had pierced the skin had gone down the muzzle of the animal and, presumably entering a nostril, had reversed course to penetrate a sinus cavity and somehow inflict death.

Nothing was found to account for the mortal injury, yet whatever it was left a channel about two fingers wide through the meat of the head and caused a severe bloodshot condition. Strangely, the brain was undamaged.

There were other puzzles. One was that the Chomiaks' bedroom window directly overlooked the spot where all this occurred, yet they saw or heard nothing unusual. Then there was the opinion of the woman veterinarian consulted by the police. The case reminded her of an animal struck by lightning, yet the night had been clear and peaceful.

Finally, the Chomiaks had experienced another mystery a few weeks earlier when one of their cows disappeared. Although well fenced in, it had vanished without a trace, and a careful search of the entire farm produced nothing. Nor were the neighbors able to help.

By now it was apparent that the cattle mutilators, whoever they were, had entered the Playground of Gods and that they were on the trail of the "meteor" even though that extraordinary object was not supposed to have been seen so far west.

From the Playground the trail of sightings was picked up again in southern Alberta. There the presence of the meteor had been officially recognized and the mutilators appeared in force when their time came, although for some reason that time seemed to lag behind their appearance in the U.S. by as much as three years.

In 1979, when mutilation problems across the border had reached such a point that the problem had become a federal issue, with Senator

Harrison Schmitt of New Mexico calling for a full investigation, the RCMP was just starting to take a serious look at it. Yet even at that time there was little real excitement, and the idea of alien involvement was firmly discounted (as it was to a lesser degree in the States).

"We're looking for humans doing these things," said Corporal Lyn Lauber, head of the inquiry in Alberta, "not UFOs or little green men."

So the difference in the mutilators' timing between the two countries may have been more apparent than real. Possibly, because of a down-play in police and reporting methods, they had entered Canada well before they were publicly recognized as anything that could become a major mystery.

Whatever the case, they made it known after a while that their mission was the same on both sides of the border, which was — and is — to cut up animals, mostly cattle, and remove certain parts.

Since the nature of the parts may help explain the purpose of this macabre branch of UFO operations, and since we are for the moment considering the Canadian part of the problem, let's see what the aliens were after during a few typical days in Alberta in the fall of 1979 when the action was well underway. Remember at the same time that the booty then was similar to what is still being taken in the western regions of both countries. Later we shall try to decide if the aliens, assuming they are responsible, have a useful motive in all this or are simply being playfully unpleasant.

As reported in *STIGMATA* (first quarter — 1980), a periodical entirely concerned with the mutilations problem which is published in Paris, Texas, by Tom Adams, here is a group of incidents during a few fall days on ranches in Alberta:

Sept. 21, SW of Claresholm — bull — genitals, tongue, tip of right ear and right leg removed.

Sept. 23, S of Brooks — five-month-old female colt — mammary glands, groin skin, right eye, an ear and part of muzzle removed.

Sept. 26, SW of Nanton — three-year-old Hereford bull — scrotum, right ear and part of tongue missing.

Sept. 29, Gem — six-year-old Hereford bull — penis missing, line of punctures along brisket.

Sept. 29, W of Kipp — yearling heifer — tongue, right ear, part of rectum gone.

Sept. 30, Parkland — steer — rectum, tip of both ears removed.

Oct. 1st week, near Elk Island National Park — three-year-old bull — genitals removed — lay with head folded under carcass.

In a moment we shall turn south into the U.S. to examine the mutilators' scope of action which, though covering at least half the country by this time, still seemed to be centered along the original meteor trail for its most dramatic events. But first we should continue our trip farther east into Saskatchewan where, it will be remembered, a shining metallic-looking ball was seen over the southern area around Gull Lake close to Alberta on the same eventful afternoon in 1972. It was almost like a sign that events there would be an offshoot of the main activity farther west, and so they appeared to be.

When their time came, seemingly a few years behind the U.S. incidents as was the case elsewhere in Canada, mutilations in Saskatchewan were a comparatively small problem. Yet all the same professional-looking evidence was there and so, too, was something else in a unique case that visibly contained what was perhaps a key to the aliens' methods.

But first, let's single out a mutilation case in Saskatchewan more significant than others in that traces were left of what apparently was a flying device. The incident, involving the death of a young steer, occurred in mid-October, 1979, on a farm in the Melville area in the southeast corner of the province.

"It was lying there in the ditch and I couldn't figure how it got there," the owner told a reporter from the Melville *Advance* (Oct. 24, 1979). "I checked and all my fences were okay. But the animal was just lying there in the ditch, with a leg broken off. . .no blood anywhere, either on the animal or in the ditch. I thought then that it had been hit by a truck or a car. I noticed an odd semi-circular track going into the ditch and then out of the ditch, where the steer had been dumped. But I also noticed that the track did not show any tire imprint. It was strange, really strange.

"We hauled the steer out of the ditch and back to the farmyard and it was then that I noticed something else about the animal. The tongue was missing and one ear had been cut off, the ear in which had been implanted an ear tag for identification purposes. Then I noticed that the animal's throat had been cut. . .very neatly with something that had to be extremely sharp. The thing is there was no blood anywhere, either on the animal or in the ditch."

Already in that first part of his account the farmer, presumably unknown to him, was giving details similar to those in mutilation cases ranging up to thousands of miles from Melville, Saskatchewan. The absence of blood whether drained from the animal or not, the neat cutting, the removal of tongue and ear (mention of the ear tag raised a question of whether the mutilators kept some kind of gory record of their surgery), the one injured leg, even the part about the ditch, all carried the mark of the mutilators — airborne mutilators, that is.

About the strange track the farmer went on to say:

"Whatever made that track had to be pretty heavy and there was something else funny about it. In the dirt by the side of the road the truck made an indentation, but when you project where the truck went into the ditch through some weeds, the weeds are not broken or damaged in any way. It's just as if whatever dropped the steer there was hanging low over the ground. I just don't know what to make of it."

Not having seen anything unusual in the area recently, the farmer thought the animal's death must have occurred the previous weekend when he and his wife were away and when others nearby did see something unusual.

"I was talking with some guys from Little Black Bear this week (Little Black Bear Reserve) and they told me that they had seen something strange one evening during the weekend."

"What was that?"

"Well, they were out during the night, driving along a country road on the reserve and noticed three or four silver-colored objects, quite large, hovering above the ground nearby. When they flashed their lights at the objects, the objects took off in a hurry. Maybe there's a connection, I don't know."

It was possible, however, the farmer did know that during another autumn just five years earlier, there was another sighting of flying silvery objects in the same area of Saskatchewan that was so remarkable it was carried as a wire-service news story. Even so, there was still less chance of his seeing a connection between that incident and the death of his steer, as animal mutilations of this sort were virtually unknown in Canada in 1974 (and, as we have seen, had barely started in the U.S.). Moreover there was never any report that the 1974 incident involved death or other loss of any cattle. It was only after a careful re-run of a tape recording of the incident that I discovered there was indeed a clue which seemed to connect that case to what we now know about animal mutilations. It was an eye-witness description of a device conceivably used in these cases. The reference was brief because obviously at the time of recording there was no point in trying to elaborate on the gadget. But put into the context of the mutilation story as it later developed, it comes through with impact.

The tape was made by D.F. Clausen and D.W. Coulthard of Regina, Sask., who were Canadian pioneers in their study of the UFO mystery. At what must have been no small expense to themselves, these two taped a long on-the-spot interview with the principal witness at his farm, made other recordings about the case, and sent the whole package without charge to other ufologists, of which I was one as publisher of *Canadian UFO Report*.

To use the term that became popular later, the case was a close encounter. It occurred one cold September morning with ice crystals falling when Edwin Fuhr, a rapeseed grower, almost drove his swathing machine into a spinning dome-shaped object on his farm at Langenburg, close to the south end of the border with Manitoba.

Without taking in all the details at first and thinking it might be "one of those new goose blinds" used by hunters, Fuhr dismounted and cautiously approached the object on foot until an inner voice told him to "keep your distance." Frightened now, Fuhr moved carefully back to his swather only to discover with shock that there were four more objects spinning in the grass on the other side.

Either paralyzed by fear or held by some nameless force, Fuhr found himself unable to move the controls of his vehicle and sat there for about 15 minutes, watching the top-like discs and feeling dismayed because he "didn't know what else to do." It was then he noticed something more was happening. His description was not entirely clear but it seemed as if an extension of some sort appeared from one of the group of four.

"It looked like something odd was poking out of it," he said, "something that was poking around in the grass. It was like a probe about the size of a 50-cent piece jumping around. Afterward I could see the grass was all twisted and there were marks like something had jumped here and there. It looked like it was probing. Anyway, that's what I thought. That went on for about 15 minutes."

Listening to Fuhr's description of the "probe" moving about gives one a weird impression the objects were occupied, and this feeling is strengthened by his account of their departure. He said that suddenly, as if by signal, they shot upward one after another in "step" formation. At about 200 feet, still in formation, they hovered for a moment before climbing quickly into the low overcast. Later, when he had overcome his shock, Fuhr told his family the circular thing had a shine like brushed stainless steel, was about five feet in height, 11 feet in diameter, and it hovered just above the ground where it was partly hidden by the tall grass of his crop. He thought it was the movement of grass around the twirling object more than the object itself that first caught his attention.

Enlarging on that picture in our minds and knowing something now about our visitors, we might visualize one or two small humanoids lying prone in each craft to perform these maneuvers, or perhaps here is a function for micronauts, a name given to occupants of alleged tiny UFOs (esp. reports from Ahmad Jamaludin, Pahang, Malaysia). A full crew of them could readily fit inside such an object, which may have an important function in mutilation activities as we shall see.

(In a later book I plan to discuss the curious subject of micronauts and their tiny craft. But having told one young witness her case would appear in this book, I will summarize it now.

(In March, 1980, Linda Crighton of Mill Bay, B.C., and several school friends reported seeing an illuminated object about two feet long flying at tree-top level. After a moment it carefully lowered itself through a narrow space between branches until it was close to the ground, where it hovered. As some of the children stepped nervously forward, the object moved away and vanished. To the teen-agers its behavior suggested awareness and control.

(A noteworthy point is that an illustration of an object matching this one in shape and lighting, though of much larger size, appears in the U.S. Air Force's Project Blue Book no. 14 — 4th edition, p. 80).

Although Fuhr's case was a single-witness one, it was impressive by any other standard in ufology. Among the factors making it so were these:

Five unexplainable circles, with the grass pressed flat in a clockwise direction, were left in the field where Fuhr said he saw the disc.

Later two more similar circles appeared nearby. These were

examined by police and photographed before spectators spoiled the evidence. No wheel tracks were in the vicinity except those of Fuhr's swather, and the circles did not look like anything a whirlwind could create.

Animals in the vicinity were unaccountably excited the night before Fuhr's sighting, with dogs barking and livestock bawling. Cattle in an adjoining field broke their fence in three places.

Another sighting of a similar object was reported in the same area about the same time.

Fuhr's memory of events at the time seems to have been interrupted, an experience sometimes associated with UFO close encounters. He could not remember certain details of operating his swather when he first saw the disc, and his inability to move the vehicle seconds later also fits into the picture of such an experience.

Regardless of its rating, however, the Edwin Fuhr case has a place in our study mainly because of a small point that is almost lost in the taped dialogue of the report. It comes out in this exchange between Fuhr and his two interviewers:

Q: Since this thing became publicly known have any of your neighbours told you that they have seen anything?

A: Yes, my neighbour here, he's got markings on his cow.

Q: What kind of markings?

A: He told me the other day there were markings on his cow just like a chicken's prints, like a Y and a cross with a straight line through the center of it.

Q: How big?

A: I don't know. He just said it looked like a chicken had jumped up the side of his cow.

Q: It wasn't a brand, eh?

A: He doesn't know. So I said maybe the cow was in the bush and got scratched.

It is hard to believe that anything as commonplace as a branding or a scratch could have caused even this very minor mystery in Saskatchewan farmland. But it was possible that the neighbour did not explain his reason for mentioning it very well and that, by the same token, the mystery was not minor at all. For later on, in what seems to have been a reversal of the usual Canada-U.S. pattern, American cattlemen who were already well into the mutilation problem were to be introduced to marking puzzles of their own.

For more on this let's return along the meteor trail and follow it far south across the border into New Mexico, a state adjoining Utah, over which the so-called fireball was first reported. By the mid-70's cattle mutilations were quickly becoming an unpleasant fact of ranching life along the U.S. section of the trail. However, it was not until the summer of 1978 that a rancher in New Mexico hit upon the idea of a test to see if the mutilators chose their victims ahead of time.

At a UFO conference in 1979 ("UFO — '79", San Diego, CA, Nov. 17) Jim Lorenzen, International Director of APRO, described how this came about. He said:

"An area of heavy activity (in mutilations) is northern New Mexico and southern Colorado. On one ranch in particular, over a period of a few years, six or seven head have been lost in this manner. Dr. Howard Burgess, an APRO member studying this situation, noted that 90 per cent of the mutilated animals fell into a particular profile — they were four and five-year-old cows, and heifers less than one year old. Nearly all were of the Charolais/Hereford strains. The cattle that were molested in this area showed signs of having been hauled aloft by a line around one leg. The hair would be worn off in a band, and the leg chafed and usually broken. In many cases they showed evidence of having been dropped from considerable height, in that the carcass would contain a number of broken bones.

"Dr. Burgess theorized that because the cattle were grabbed at night without warning, but still seemingly carefully chosen, they were somehow marked in advance.

"He convinced a local rancher to run a number of cattle under a 'black light' at night. They found another 20 cows, all of which fitted the profile of those previously killed, marked in a manner that became luminescent under the black light. The cause of this luminescence was never determined, although many tests were run in the attempt to find out.

"There is a marked difference between the mutilated cattle and those dying of natural causes. The 'mutes' do not bloat and scavenging animals do not go near them. They usually exhibit a hole in both jugulars, apparently where the blood has been drained. Even early on, the internal organs are abnormally mushy, and some medical technicians who handled these organs in the course of autopsies have complained of health problems."

A conspicuous feature of the markings examined during the black-light test was that they were on the animals' backs or high up the sides where they could be seen from the air, if that was the purpose. In that respect there was a similarity to the markings in Saskatchewan which, as we gather from farmer Fuhr's neighbour, were likewise on his cow's back or close to it.

There, however, the similarity seemed to end. In New Mexico an organic material of some sort was used, whereas in Saskatchewan the marks were evidently scratched into the cow's hide and there was nothing to indicate they were visible at night. But I do not think we can argue there was no possible connection because of that.

In the first place there was a span of up to four years between the times that the marks in the two cases were made. In that period the aliens, if they were responsible, might well have changed their technique to help them in the greater security of night operations.

Secondly, there was unexplained aerial activity in both areas when the animal marking might have been done. On the Canadian side, Edwin Fuhr had a rare good look at strange flying vehicles which were made all the more curious by a gadget that jumped about in the grass. Was this a device for marking cattle? Perhaps at the moment of observation it was being tested for some reason in the grassy hideout when Fuhr came upon the scene by chance. Later we shall consider an example of how it might have been used.

Over cattle country on the U.S. side there were equally strange and apparently more numerous events in the air that gave the same impression of an alien intelligence at work, as we shall now see.

No matter how little connection one may decide there was between the "meteor" of 1972 and the UFO "invasion" of 1973 that led into the problem of cattle mutilations, it surely is at least of passing interest to note that the former was first observed over a point close to the area where the attack against cattle became most troublesome. That area was a section of mountainous country mostly in northern Colorado, next door to Utah, and if ever there was a place that might draw alien cattle hunters and at the same time cause them to sniff about before coming in for the kill, that would be it. For close to cattle ranches in some cases are military bases utilizing nuclear energy which, unexpectedly, may attract the sky visitors rather than cause them to steer clear.

As part of a series of articles on the UFO mystery carried in the *National Enquirer* for which he was a reporter at the time, Bob Pratt gathered the opinions of UFO-watchers who had observed this relationship (and here a word of thanks to Bob for sending me his typing copies to ensure I had a complete set of his reports — Sept. 6, 1977 to May 30, 1979 — on the mutilation problem). In one article he referred to his talk with Todd Zechel, former director of research for Ground Saucer Watch, a national UFO organization, and a former member of the U.S. Army Security Agency, who told him:

"I think there is a strong possibility that UFOs are doing something to replenish themselves around the military bases. There's got to be some reason why the UFOs are repeatedly seen over and near the U.S. military bases other than just chance.

"I think it is interesting that in the reports that I've seen released by the U.S. Government, a number of them concluded that UFOs were demonstrating a clear intent in the nuclear weapon storage areas. There is information in the underground network from military sources indicating nuclear material has been taken during UFO penetrations of the bases (in 1975).

"I don't think they would have any use for nuclear material as weapons but there seems a possibility they would want to use this for fuel. I find it very startling, if not frightening, that these sorts of things are going on and the government isn't more concerned about it.

"During the past month I have investigated cattle mutilations, and after talking with people in New Mexico and Colorado — ranchers who have had mutilations, police officers and scientists who've investigated — considering all the evidence at this point, there's no doubt in my mind that cattle mutilations are not only related to UFOs but that mutilations are part of an extraterrestrial experimentation program.

"The preponderance of evidence certainly leads me to conclude that the mutilations are something in the order of an implantation of an extraterrestrial organism, in which the extraterrestrials are attempting to find out how this organism will interact in the earth's environment."

As it happened, possible support for this idea was provided when carcasses with the strange markings were examined in 1978 (as described in Jim Lorenzen's earlier quote) at the Schoenfield Clinical Laboratories in Albuquerque, New Mexico. Finding that the "organic material" on the hides contained far more potassium than the normal amount for earth samples in that part of the country, Bob Schoenfield further discovered that whole areas of tissue were missing from inside the carcasses. He also found that some sort of cylindrical object had been inserted into one animal which evidently had burst, completely shattering the pelvic bone of the hapless creature (as reported in *The Night Mutilators* by Gene Duplantier).

It sounded like the result of a cruel prank but apparently Schoenfield thought otherwise.

"A lot of people think it's just some kind of hoax, but something like this really makes you sit back and wonder," he said. Unfortunately it seems nothing more was said about that particular mystery and we are left to speculate that an alien testing device in the animal exploded by mistake.

What may have been another incident involving something implanted in an animal occurred in 1977 on a ranch near Sterling in northeastern Colorado. A neighbouring farmer, David Wagner, told Bob Pratt what he saw.

"I had a calf missing and there's a pasture a quarter of a mile from me and I went over to check on this calf. I saw this cow laying there by a gate. I didn't pay much attention to it. It looked like she was trying to have a calf and I thought they had a vet there and figured she died so they just let her lay there.

"They had a piece of paper about 18 inches wide and 3 feet long laying right under her rectum. I don't know what kind of paper it was, some kind of waxed paper it looked like."

That was about 4 o'clock in the afternoon, Pratt wrote, and according to Wagner nothing had been done to the cow except for possible cuts in the rectal area. Wagner went back about the same time the next afternoon, discovered the cow had been mutilated during the intervening time and he notified the owner. The paper was gone.

"They took the tongue out, cut one tit off, one ear off and had the jawbone cleaned off just slicker than a whistle," Wagner said. "And they had — it was just as slick around the rectum. . .about 9 inches around, and pulled that all out."

Wagner said that on the previous afternoon the pasture's owner had been there until 2:30 and saw nothing unusual, which meant whatever was done to the cow to kill it was done between 2:30 and 4:00 p.m. So here we have a striking example of the speed with which the mutilators operate (suggesting again their use of aircraft of some kind), their alert anticipation of human approach and their uncanny skill with their tools.

On that last point we have the words of the veterinarian, Dr. L.L. Rieke, who did an autopsy on the cow.

"You could assume that I don't entirely go along with what you call mutilations," he told Pratt, but he admitted he was puzzled by this case.

"There's something happened there that I don't know for sure about. When I posted her, the uterus was gone, taken out from behind. . .which was unusual."

Asked how he could explain the missing uterus, Dr. Rieke said: "I don't. . .that's still got me baffled. I don't know."

In this case we also have an apparent search for something so special it seems unlikely it was a natural part of the animal. The paper laid under the cow, which was unusually plain evidence for the mutilators to leave behind, and removal of the uterus in some ingenious manner, suggested there was something here that required very careful handling. A delicate register of some kind could have been the answer.

While cases like this do at least hint that the aliens, if they are indeed responsible, are using a device that will help them record conditions of our environment, they still fail to answer a major question in the mutilation puzzle. If the mutilators are so concerned about environment, why all the fuss about various animal parts other than certain internal organs? What, for instance, is the connection between atmospheric effects and a cow's ear?

Part of the answer may be that here we find our UFO visitors doing something they like best, which is playing tricks on us, presumably to keep their identity and purpose in all things as secret as possible. At least two reputable students of the UFO mystery, John Keel and Dr. Jacques Vallee, have written at length on the subject, giving strong reason to believe that fakery is high on the UFO agenda.

As masters of the art, our visitors may be putting their skills to good use by making it appear that cult-minded humans are responsible for the mischief, although they over-reach themselves in the extravagance of their performance. As hard-nosed investigators admit, no cult on earth is known to have the funds and skill for what the mutilators do, and no human criminals are so adept at avoiding the law. To date, after thousands of mutilation incidents, not a single arrest has been made in those cases which are not simply copy-cat efforts. Even so, mutilating with such wild abandon may be part of the aliens' act to keep the search alive (though reported cases are fewer now) for crazy human culprits.

I suspect, however, that an answer may also lie somewhere in a discussion I heard at MUFON's 11th annual UFO symposium in Clear Lake City, Texas, during June, 1980. The talk centered on a remarkable documentary film called *A Strange Harvest*, prepared and produced by Linda Moulton Howe, director of special projects for a TV station at Denver, Colorado.

The "strange harvest" was that of cattle mutilators, and a view was expressed at the meeting that their collection of parts might not be as frivolous as it seemed. Instead the speaker, a specialist in necrobiology and anatomy, saw a certain expertise in the mutilators' choice of booty. He said extracts from what they took (like that sample of a few days' work in Alberta) could well be used to concoct a serum of medicinal value that might be far above human standards if UFO creatures are indeed responsible. Purposes and beneficiaries of the serum would, of course, be unknown.

In their hunt for animal parts the mutilators proceed with vigor, yet somehow with the seeming alien flair for mystery, they remain phantoms to us while being terrifyingly real to their victims. Investigators must often settle for tracks in deciding whether any outsider was present at all, and the tracks usually present a mystery. In the area under study a type sometimes found is a triangle of almost circular imprints, four inches in diameter, on sides measuring 28 inches. Investigators call them pod marks, and here we are reminded of tracks left by the Beast of Devonshire (Chapter Three) which, though more oval shaped, also had a measurement of four inches. In mutilation cases a triangle seems to have replaced the Beast's single-file pattern but it could be we are looking at something of the same origin. However, for cattle work the pod apparently is on the undercarriage of an agile flying machine rather than on the rim of a floating wheel.

In an account given to Bob Pratt one can picture this machine in action quite clearly after tracks showed it left another machine which stood on a larger triangle of pods measuring 14 inches in diameter and placed about six feet apart. The scene was a ranch owned by Manuel Gomez on an Apache reservation near Dulce, New Mexico, in the heart of mutilation country. When the first mysterious death of a cow occurred there in June, 1976, State Trooper Gabe Valdez told Pratt:

> We could trace where the animal was eating at the time they chased her. They chased the cow a good 600 yards, and they hit her with something. We suspect some kind of a tranquilizer. She fell down, walked another ways, fell down, then got up again and then she finally fell down and died.
>
> We could tell all this from the marks in the ground. The ground was real soft. . .We could tell she'd been chased because you could see the small tripod prints after her and where the cow was running away.

> There were two or three of the little objects following the cow. Those tripod marks were all over the field. They chased the cow and they returned to the craft. (Valdez earlier called it a "big aircraft" because of its large triangle of heavily indented pod marks.) We were able to follow the tracks all the way to where the cow was mutilated and back to the aircraft again.
>
> Every place the small tripod followed the cow it would scorch the grass. We could see the trail where the grass was scorched.
>
> We also found an oily substance right in the middle of these small tripod marks. But we couldn't analyze it because it disintegrated. All we know is the laboratory said it was a petroleum-type substance. They couldn't identify it.

This particular case does much to clarify our picture of the mutilators' personality and equipment, even though they are still mostly phantom. We already know from the case of the dead cow lying on "waxed paper" how quick they are, but here they worked at stop-watch speed.

When the cow's owner, Gomez, discovered the dead animal, its mutilation looked about half finished. "Some of the stuff that had been removed was lying alongside of her," an investigator said.

Jumping into his truck, Gomez rushed to town 12 miles away to fetch Valdez. When he returned with him, the surgery was neatly finished, the parts were gone and there were pod marks on his departing tire tracks to tell the rest of the story. The only marks near the cow's carcass were more pod marks.

Although I have never seen it spelled out, the culprits' speed of escape in these cases may help explain why natural predators are usually blamed despite the clean, sharp incisions so often observed. After a year of investigation financed by the state of New Mexico, former FBI agent Kenneth Rommel, Jr. released a report in 1979 which said, in effect, that natural predators were responsible.

What neither Rommel nor anyone else knew at the time was that the culprits, besides being skilful, were persistent to an extraordinary degree. Three years after that report, Gomez, whose ranch was a favorite target for some reason, was still being plagued by the same kind of mutilator which to that point had claimed a total of nine of his animals. As there seemed to be no safeguard that could check the killings, his cattle insurance was cancelled and he was forced to turn to another company at a higher premium.

Yet despite their most careful vigilance, the only unusual activity seen by the Gomez family and their neighbors in all that time were strange lights of varying colors moving silently about overhead.

"They like to stay up in the trees close to the mountains where radar can't pick them up," Gomez' son told the Albuquerque *Journal* (May 31, 1982), to which his father added, "Mr. Rommel never did come out here to talk to me."

Thus we are still left with the complex mystery of a predator that leaves triangular tracks, or one that can toss around an animal weighing close to a ton, or one that performs surgery quickly on an open field that is impossible for a human to do with all the latest equipment, or one whose operations have unexplainable effects. Here are examples:

In a case almost identical to that of the mutilated steer found in a Saskatchewan drainage ditch 700 miles to the north, with no tracks evident, a rancher in Colorado discovered his mutilated heifer in a drainage ditch with no tracks around and no dirt on its feet that stuck straight up. Yet the ditch was full of mud.

Because of many broken bones, some mutilated animals appear to have been dropped from a considerable height. It is thought they were hauled aloft by a line attached to one leg, to judge from chafing marks around the leg.

An investigator in Colorado reported the case of a carcass found wedged in the crook of a tree.

In some instances the mutilators remove patches of hide without cutting the hair that lies flat. "It's as if the hair was standing up and the blade went right between the hairs. . .which is impossible to do," a sheriff said.

Working in laboratory conditions with laser equipment for Linda Howe's *A Strange Harvest*, technicians tried without success to duplicate the skill shown in a mutilation case.

An operation on one animal that involved removal of the anus and uterus was humanly impossible, according to the vet who examined it. "There's no way a man could get two hands up and in to cut those tubes off, just straight across," he said.

In New Mexico a rancher collected insurance on a mutilated cow even though the operation had changed the cow's color from red to cream.

While investigators seem sure, in public anyway, that the mutilators are all creatures of an earthly kind, the evidence discussed so far and in the next chapter may also point to a scenario something like this:

Creatures that resemble humans in a general way but are smaller in stature, arrive in their flying machines over a ranch at night. Some of the machines are disc-shaped and measure just a few feet in diameter, while two or three others are box-shaped and much larger, giving an unlikely observer the impression they are helicopters. They are painted black as if to improve their chance of being unseen or at least more difficult to describe.

The disc-shaped vehicles fly low over the cattle as the occupants, operating an arm extending from the craft, manipulate it to make a mark on the backs of some of the herd that catch their eyes for some reason. Meanwhile the "helicopters" land beside the herd on their tripod undercarriage and discharge a clutch of smaller craft of similar design. These do the "cowboy" work of the squadron. Flying almost at ground level, with a scorching jet coming from their underside that gives them tremendous acceleration, they chase a few of the herd with knockout guns.

When an animal falls, small occupants of the attacking craft, armed with surgical tools, spring out (or more likely in true UFO fashion they float out leaving no footprints) and dismember it. With their typically thin arms and long fingers they can reach inside their victim and remove its organs with remarkable dexterity. They take these parts back to their craft.

Some of the animals they attack have been marked in previous encounters. From these the humanoids remove a cylindrical device which they carefully wrap and take away. Animals newly marked by their colleagues in the discs are knocked down by a "cowboy" crew in similar fashion but without serious injury. These are implanted with the same sort of capsule for later removal and are then released.

There are other exceptions, as in some cases it is more convenient to operate on an animal inside one of the vehicles. For this purpose the victim is hauled aboard one of the large craft by a line around one of its legs, if the craft is airborne, or by a light beam as described in the next chapter. When the surgery is over the dead animal is pushed out. Sometimes it seems a tidy way of doing things to drop the carcass in a drainage ditch.

When all the field work is finished, the small craft return to the large vehicle where everything is taken aboard. The mutilators then depart.

Chapter Nine: Episode In "A Strange Harvest"

Now that we have learned something about the clues left by the cattle mutilators, let's see if somehow we can have a close-up look at the creature who is capable of such an unpleasant yet clever performance. Because he is supernaturally quick and cunning, the chance of having any look at all seems virtually hopeless. Thanks, however, to that extraordinary film *A Strange Harvest*, there is on record a close-encounter case so vividly related it is almost like being there.

Writer, producer and director of the film was Linda Moulton Howe of KMGH-TV in Denver, Colorado, who was spurred on by the belief that the mutilation story will go down in history as the greatest mystery of the 20th century. Also she believed that hers was the state where it all started when, in September 1967 on the King Ranch in the San Luis Valley, an Apaloosa mare named Snippy was found dead and stripped from the neck up. Some said then it was the work of flying saucers. To date no one has been able to prove otherwise.

To make the 90-minute documentary, Linda spent seven months on research during which time she and a camera unit travelled throughout Colorado, Wyoming and New Mexico. They interviewed journalists, law enforcement officers, ranchers and scientists. Having been told that only crews working from helicopters could do that kind of mischief, they went to an air force base to check out that possibility. There they concluded that a chopper large enough to handle a cow and the equipment necessary to cut it up would make so much noise that an animal would not stand still for its approach and landing.

For the scene that follows as a segment of the film, all the notes and conversation of the original transcript are kept intact to help recreate the studio conditions in which the subject, Judy Doraty, recalled her experience. In a state of hypnotic regression she talked to Dr. R. Leo Sprinkle, a psychologist and director of counselling and testing at the University of Wyoming.

This portion of the transcript is reproduced with the kind permission of Linda Howe and station KMGH-TV.

FADE UP ON: DR. LEO SPRINKLE IN HYPNOSIS SESSION WITH JUDY DORATY.

SUPER FAR LEFT OF SCREEN: PROF. LEO SPRINKLE, Director Div. of Counseling & Testing University of Wyoming.

SUPER FAR RIGHT OF SCREEN: JUDY DORATY

SPRINKLE TALKING UNDER NARRATION:

As you maintain this deep concentration, this deep relaxation, you can feel yourself even going deeper yet. And as you go deeper and deeper, feeling comfortable and relaxed, allow yourself to go back in time, going back and back, back in this time of March 1980 to the year 1973, to that particular night, the night in question in which you and your family members had a sighting that was an unusual event. You'll be able to go back in time, you'll be able to talk and describe your impressions. Now let yourself drift back, going back from 1980 to 1979 to 1978, 77, 76, 75, getting closer and closer to that particular time, that particular place. In 75 to 74 and now 73.

HOWE:

WHAT YOU ARE ABOUT TO HEAR IS PERHAPS THE FIRST AND ONLY KNOWN EYEWITNESS ACCOUNT OF A MUTILATION. BECAUSE DR. LEO SPRINKLE IS RECOGNIZED AS ONE OF THE WORLD'S LEADING

AUTHORITIES ON CASES INVOLVING UFOS AND HUMAN ABDUCTION, HE WAS CONTACTED BY THE AERIAL PHENOMENON RESEARCH ORGANIZATION IN TUSCON, ARIZONA, TO DO HYPNOSIS WORK WITH THIS WOMAN. HER NAME IS JUDY DORATY. ON A CLEAR MOONLIT NIGHT IN MAY 1973, JUDY WAS DRIVING BACK FROM A BINGO GAME IN HOUSTON, TEXAS. SHE AND HER MOTHER, DAUGHTER CINDY AND BROTHER AND SISTER-IN-LAW WATCHED A STRANGE BRIGHT LIGHT HOVER IN THE SKY. AFTER THAT NIGHT SHE SUFFERED TERRIBLE HEADACHES AND ANXIETY.

IN 1978, SHE UNDERWENT ONE HYPNOSIS SESSION WHICH BROUGHT OUT PART OF HER UNUSUAL EXPERIENCE — BUT NOT ALL OF IT. SHE ALSO HAD SEVERAL PSYCHIATRIC AND PERSONALITY EVALUATIONS ALL WITH RESULTS IN THE NORMAL HUMAN BEHAVIOR RANGE.

In that year, now just letting yourself go to that particular month, that particular night in which you were driving along and you had the sighting of something unusual in the sky.

OK, Judy, maintain this deep concentration, this deep relaxation. Now just let yourself drift back. You're there driving along, you stop the car. You get out. You look. Feel yourself there. Now you're standing Outside your car. Now you're looking up. You can describe your impressions. What are your impressions?

THIS SECOND HYPNOSIS SESSION WITH DR. SPRINKLE OCCURRED ON MARCH 13, 1980.

DORATY: It's like a spotlight shining down on the back of my car. And it's like it has substance to it. I can see an animal being taken up in this. I can see it squirming and trying to get free. And it's like it's being sucked up. It's taken into some sort of chamber. It's a little round, tiny room. And I get nauseated at watching how they excise parts. It's done very quickly, but the calf doesn't die immediately. For some reason, the calf's heart isn't taken. . .I don't know, it seems like it's still living and that upset me very much. (She sighs). And then I can see the calf being lowered. . .like it's being dropped back down and when it's on the ground it's dead. I can see that it's not moving. But I feel very sick to my stomach at what I witnessed.

YOU SEE THE ANIMAL BEING CUT UP?

Yes.

HOW IS IT BEING CUT UP?

Dissected.

WITH KNIVES?

Instruments of some kind.

CAN YOU DESCRIBE THE INSTRUMENTS?

They're like a knife, but they have different kind of handles. More like a razor. It doesn't have movable parts, but like a straight razor type.
There's tissue. . .it's laid out flat and smooth. It glistens. And there's, there's needles in it, or what appears to be like needles. It may be probes, I don't know. But it has a tube connected to it. And the same thing with what appears to be testicles. I can't tell. There's a probe or needle or something inserted and the tube leading. . .The same with the eye. And the tongue. The tongue is cut, but it's not cut lengthwise, but this way, up and down.
And I feel like I'm not supposed to talk about it really because it was like a warning went off when I observed this.

A FEELING THAT SOMEHOW YOU'RE NOT SUPPOSED TO TALK ABOUT IT?

Right. I'm not supposed to. I shouldn't have observed it.

LIKE YOU'RE WATCHING SOMETHING THAT'S. . .

I was told mentally that it was an accident, I should not have seen this.

IS ANYBODY ELSE WATCHING WITH YOU?

No.

IS ANYBODY ELSE AROUND YOU?

I feel the presence of other things, but I don't know what they are. I can't see them.

YOU FEEL YOU'RE STANDING BY THE SIDE OF THE ROAD. . .BY YOUR CAR NOW? OR ARE YOU STANDING SOME PLACE ELSE NOW?

I feel I'm in two places at once.

OK, DESCRIBE HOW YOU FEEL.

I feel like I was able to see what was going on in the craft, but I was also standing beside the road.

OK, JUST LET YOURSELF FOCUS ON WHAT'S HAPPENING INSIDE THE CRAFT. BE AWARE OF WHO IS AROUND YOU. IT'S OK. YOU'LL BE ABLE TO DESCRIBE WHO IS STANDING AROUND YOU.

(She pauses)

It appears to be two little men.

TWO LITTLE MEN? OK, HOW DO THEY LOOK?

They look. . .their hands look funny. They have long, claw nails like. . .

They have very large eyes. They're very hypnotic. Like they're so big that . . .and they don't blink. The eyes do not blink. It's almost like, I guess. . .a snake?

They were very snappy about their movements and they knew exactly what they were doing and I felt a little better because I, for some reason they projected that it was necessary that this be done . . . you know, that it was for our betterment, for the betterment of mankind that this was done . . . that they were more or less watching out for us.

SO YOU FELT BETTER WHEN YOU HAD THAT IMPRESSION?

Yes.

NOT ONLY OF PRECISION, BUT ALSO A BENEVOLENT OR KIND ATTITUDE TOWARD PEOPLE?

Right . . . I can't say that they were kind.

CONCERNED?

I can't even say that they were concerned. They were just like doing a job that. . .evidently it affects them, too.

And it has to do with somehow nuclear waste or testing or. . .

They've been here for quite some time and they test the soil as well as our water as well as our animal life, our vegetation. I questioned why they aren't saying more and why they can't stop it if they're so knowledgable and they get angry. I don't know. I don't understand that.

YOU DON'T KNOW WHY THEY GET ANGRY?

I don't know why, I don't know why they get angry and I don't know why if they can't stop it, if they're so knowledgable.

HOWE:

AT ONE POINT IN THE HYPNOSIS SESSION, JUDY DOESN'T UNDERSTAND WHY SHE CAN SEE HER DAUGHTER, CINDY, IN THE CAR ONE MINUTE AND GONE THE NEXT.

SPRINKLE: BREATHING DEEP AND EASY, LET YOURSELF RELAX. YOU CAN BE AWARE OF YOURSELF STANDING BESIDE THE CAR. YOU CAN BE AWARE OF YOURSELF STANDING IN THE CRAFT. NOW LET YOURSELF BE AWARE WHERE CINDY IS.

I know I have a feeling of fear.

OK, LET YOURSELF GO ON THROUGH THE EXPERIENCE. LET YOURSELF GO ON THROUGH THE FEAR. WHAT IMPRESSIONS COME TO MIND?

I just visualize her on a table.

OK, WHAT HAPPENS NEXT?

I'm just afraid they're going to do to her what they did to the animal. (She cries.)

JUDY STARTS POUNDING ON CHAIR ARM

LET YOURSELF RELAX DEEPLY. . .

JUDY FLINGS HER HEAD UP AND OPENS HER EYES LOOKING ANGRILY AT SPRINKLE

No, I don't want to!

YOU CAN TOLERATE. . .

Not yet. . .!

YOU KNOW IT'S ALL RIGHT. JUST LET YOURSELF GO ON THROUGH. JUST RELAX DEEPLY. YOU KNOW THAT EVERYTHING'S GOING TO BE OK. YOU SEE CINDY ON THE TABLE. WHAT HAPPENS NEXT?

I'm just afraid they're going to hurt her. (Still crying, tears streaming down her face.)

YOU'RE AFRAID, BUT YOU LET YOURSELF GO AHEAD. LET YOURSELF GO ON THROUGH THE EXPERIENCE, LET YOURSELF BE AWARE. LET YOURSELF RELAX DEEPLY. YOU CAN BE AWARE OF WHAT HAPPENS NEXT.

They put her to sleep, I guess.

OK, THEY PUT HER TO SLEEP. OK, WHAT'S HAPPENING NOW?

They're just examining her, I guess. But I'm so afraid they're going to cut her or something. (More crying.)

DO YOU SAY ANYTHING?

They don't listen, they just ignore me. . .and just go about their work as if it's nothing. They don't seem to have any emotions. They don't seem to care. I thought they were going to harm her. That they were going to do to her what they did to the animals.

BUT THEY DON'T HARM HER?

No.

NO. OK, WHAT DOES HAPPEN?

They just take some samples. . .from her.

WHAT KIND OF SAMPLES?

I don't know. Just scrapings like. . .They go inside her mouth. I can see her lying there and they just kind of scrape little pieces off.

ANY OTHER SAMPLES?

That's all I can remember.

HOWE:

THE PRECEDING SCENES WERE TAKEN FROM A FOUR HOUR HYPNOSIS SESSION. DR. SPRINKLE STATED AFTERWARD THAT JUDY DORATY IS, IN HIS PROFESSIONAL OPINION, A SINCERE, HONEST AND CREDIBLE PERSON.

WHAT HAPPENED BEFORE OUR CAMERA WAS HER FIRST COMPLETE AND VIVID RECALL OF HER 1973 EXPERIENCE IN TEXAS.

DORATY: ". . .wouldn't listen. They just went about their work. They didn't have any emotion or feeling. They couldn't seem to understand why I would even be upset.

JUST LIKE IT WAS ROUTINE. . .

Yes.

BUSINESS FOR THEM?

Like a laboratory animal or something.

Chapter Ten: Big Mama & Charlie Redstar

Lesley Footner

Unlike so many instances in ufology when we look for evidence that suggests an alien presence, there is no shortage of material on the mutilations scene. The unexplained assault on cattle in recent years leaves little doubt that a new kind of destroyer with intelligent methods is at large on western ranches and pastures of North America.

To connect the mutilations with aircraft — and more than that, aircraft of an unusual sort — the clues are less abundant but the quality is good. In the case of mutilations on the Gomez ranch we have the pod-mark evidence that aircraft were involved without question, and evidently they were aircraft of an unknown origin. If there are human cattle-rustlers who have some type of helicopter that serves as a carrier for small craft with tripod landing gear similar to its own, they have kept their equipment as secret as their identity.

In addition we have evidence that suggests cattle are airlifted and dropped at night, something any helicopter crew would find difficult to manage, and we have the early eye-witness cases of black unidentified craft.

Still, large questions remain. If airborne aliens are in fact responsible for cattle mutilations, where are the massive cargoes of meat that supposedly are vanishing off the face of the earth? For airlift work of such immensity, some kind of super carriers would be needed to take delivery of the booty as it is brought from the ground. Presumably — and again if aliens are involved — they would be conspicuously large craft more than likely capable of trips into outer space. Surely at some time during the great cattle caper, particularly when it was at its peak in the late '70s, aircraft of these dimensions would have been seen from earth. So where were they?

For a possible answer, let's meet Big Mama of the sky over Logan County in northeastern Colorado, and Charlie Redstar of the sky over Carman of southern Manitoba.

Of all the areas along the meteor trail, none had more trouble with mutilators at the height of activity than Logan County (77 unusual animal deaths in 1975-77), and for some reason the trouble was matched by a strange display of lights in the night sky.

"There is a very definite connection between the lights and the mutilations because. . .each time we've had mutilations the lights have been seen," Sheriff Harry L. "Tex" Graves told Bob Pratt.

Focal point of the display was a particularly large light, dubbed Big Mama, from which smaller ones were seen to drop before shooting away in various directions. Changing color between white and red with a hint of green at times, Big Mama would sit for a while in one spot as her offspring took off and returned. Then suddenly she would move so fast she seemed to reappear instantly in another position.

"It's too fast for this earth," a deputy sheriff remarked. "Or maybe somebody has secret weapons, whatever it is."

Big Mama and family made their debut in mid-August, 1975, and remained through the period when the mutilators were busiest.

"In 1975 and 1976 when our mutilations were really heavy we had numerous light reports coming into us almost every night," Undersheriff Jerry Wolever told Pratt. "And two or three days after that was when we had reported mutilations. . .we were receiving light calls daily and nightly and we had this one map plumb full of little colored pins, and then we started putting up our other map where the mutilations were occurring and they were in this concentrated area where the light reports came from."

Earlier in 1975 although not receiving any real attention until May, perhaps because of harsh winter weather, a light that resembled Big Mama started a similar nightly act over southern Manitoba almost 700 miles to the north. Named after its dominant color but having the same variations as Big Mama, this was Charlie Redstar whose performance became such a popular topic after its lackluster debut that crowds camped out on the fields around Carman to watch it (*Journal UFO*, vol. 2, no. 3). Like Big Mama, Charlie had a family of smaller lights that hurried in and out.

But there was a major difference. No mutilations were reported among the cattle herds of Manitoba when Charlie arrived and none were reported later, which raises this thought — perhaps Charlie

Redstar was really Big Mama, and perhaps this performance was a dry run for what would come later in Colorado.

Why the aliens would choose to practice over a Canadian site rather than over those areas in the U.S. where the real show would be staged later must remain a mystery. Perhaps it had something to do with the size of audience. Perhaps the performers wanted to keep their plans as obscure as possible and so found it more convenient to rehearse over sections of Canada where they would receive little attention. (If that was the idea, it did not work in Manitoba).

Whatever their motive, they have followed that strategy before in their cattle-mutilation plans, or so it seems in retrospect. For the evidence, let's turn to a couple of incidents in the quiet ranching district of the Cariboo in south-central B.C.

The year was 1969 when Lillian and I were there on a holiday trip, though I had a second reason in mind, which was a tip from my bank manager about UFOs in the area. He had moved from the Cariboo about two years earlier, and before we left home he gave me the names of a few former clients there who had told him about sightings.

On the face of it, the prospects were not promising. There was little industrial development in the Cariboo, which sometimes attracts our visitors possibly because of its power potential, and there was nothing in my records to show they were interested in herds of cattle which in the Cariboo were everywhere. All that made for pleasant holiday country but I could not see why the aliens would bother with it.

Then suddenly, almost at the start of our visit, I realized here was the jackpot. I was finding that all the leads on my banker's list were good ones and that most of the sightings were in or close to 1967, which was recent enough to give them good news value for my magazine. As it turned out, 1967-68 was a "flap" period for Canada and, thanks to the material I was able to gather, the Cariboo Valley topped the list nationally for a number of reports.

However, the two incidents I have mentioned are what concern us here for the way they seem to illustrate again how carefully the aliens laid their plans to attack cattle. If we have assessed things correctly, we have seen how in 1972 one of their spacecraft overflew the productive ranching country of western North America which would bear the thrust of their imminent visit. Possibly the purpose of that was just to make a giant head-count but more likely, in view of our visitors' penchant for careful planning, it included other tasks such as a test of the atmosphere. Next, while the attack was in its early stages along the U.S. meteor-strip, there were the apparent mother-ship preparations underway in an isolated area over Manitoba.

But was some of the mutilation program in progress before all of that? Did our UFO visitors, far ahead of time, go through a preliminary drill over a secluded section of western Canadian ranch country? To judge from accounts of the two following cases given to us in the Cariboo by the principal witnesses, they staged at least one full-dress rehearsal there in 1967, five years before the meteor appeared and eight years before operations were fully underway.

The setting was a likely one for such a rehearsal. Carved by the Fraser River to form a broad avenue through the mountainous center of lower British Columbia, the Cariboo valley with its ranches lies well away from heavily populated centers. Also, until its far north end reaches the sawmill smoke of Williams Lake, its air is clear. Site of the two cases we now consider lies in the heart of that country.

Case number one was that of Ernest and Shirley Hills, a pleasant couple who operate a lodge on the sparsely settled shore of beautiful Green Lake. At the time of their experience one cold October night in 1967, no one had ever heard of Charlie Redstar, which regardless of the much earlier year was unlikely anyway at this place about 1,000 miles from Carman, Manitoba, where he was to make his name. Yet what they described might well have been Charlie himself in close-up detail.

This was the report we carried in CUFOR (vol. 2, no. 4, 1969):

> "From our bedroom upstairs there is a wonderful view south across the lake," Mrs. Hills told us. "On that particular night there wasn't much to see as it was very cloudy and dark, but I was looking out when I noticed this odd red glow appear over the hill on the other side. It was growing brighter and shining off the clouds when I saw two red lights come up over the hill and then another, much larger, move up between them. Then two more small ones appeared, so there was this immense red light with four little ones — two on each side — in perfect formation around it. I had never seen a UFO before but somehow I knew right away 'this is it.' "
>
> Suddenly this was it in a still more dramatic way. Having cleared the hill, the awesome formation now started swiftly across the lake on a collision course with the lodge! Answering her frantic shout, Ernie Hills and their son, Robin, rushed in just in time to see the lights veer off on another collision course toward their neighbors, the Gammies, to the west. When this happened Mrs. Hills flew to the phone. ("As soon as Shirley Hills called, I just dropped the phone, rushed outside and stared," Bert Gammie of the Flying U ranch told us later. Unfortunately, as he and his wife were preparing to leave on a trip, we were able only to speak to him briefly on the phone during our Cariboo visit.)
>
> But at the last moment the huge glowing object with its satellites — Mrs. Hills compared their relative size to a door and a door-knob — again changed course and shot off toward the southwest end of the lake, where it stopped abruptly about seven miles away. Here it added another trick to its already remarkable performance which particularly impressed Hills.
>
> "That thing was obviously looking for a place to land but its speed was absolutely unbelievable," he said. "First it would shoot off to one side, then back again, then up and down and so on, but

moving so fast it seemed to be in several different spots all at once. Then suddenly it stopped and that was it. It had found its spot."

Because of the reflection underneath Mrs. Hills was inclined to think it was on or just above the water, but her husband believed the reflection was caused by ground mist or a thin coating of fresh snow.

When the enormously brilliant object finally came to a halt casting a glow on the clouds around it, the witnesses were reminded of a spectacular version of the B.C. Hydro building in Vancouver seen at night in the distance with all its windows alight. Although the outline of the object itself was not clear, the Hills agreed it was at least equal in size to that spacious office building.

Then, through binoculars, they made a curious discovery. Horizontally across the face of the UFO ran two parallel lines of whitish light, resembling windows lit from within and there were stranger discoveries to follow. As they watched, the four little balls of light that had maintained their formation around the large one suddenly darted off in different directions. Then there was another surprise.

"After a while those things started coming back," Ernie Hills said. "First of all a beam of light would come out of the big light, as if a door had opened, then one of those red balls would come sailing into view and shoot right down it. We weren't sure if the little lights actually entered the big one but Bert Gammie next door said he could see them going right in."

Whether each satellite made more than one trip was not entirely clear, nor were the witnesses sure they saw all four return (and we had no further word about Gammie's impressions). Puzzled, and after a while exhausted, by their unbelievable experience, the three Hills lay down in bed to watch in comfort as the drama unfolded.

But the final act was denied them. As the night wore on they all fell asleep, and when they awoke their strange visitors had left.

In CUFOR a popular feature was "Matching Cases" which described similar UFO incidents far removed in time and location. Thus it underlined the global, and perhaps universal, nature of our visitors' methods and equipment. With special help from contributor Graham Conway several almost identical pairs of cases were given. Yet I remember none more striking than this comparison between Charlie Redstar or Big Mama and what the Hills saw eight years earlier.

Case number two was that of Brian Grattan, then a young man who with his wife operated a guest ranch at Lone Butte a few miles north of the Hills' place. Just as the Hills had an experience that seemed to anticipate the aliens' use of supercraft like Charlie Redstar, so did Grattan have what might have been a preview of one of their "cowboy" vehicles in action. However, this vehicle looked larger than the spinning domes which Edwin Fuhr in Saskatchewan saw years later in his field of rapeseed and which may have "fingered" his neighbor's cow.

After explaining that in the summer and fall of 1967 there were several UFO incidents on his ranch, Grattan told us he thought the flurry was over as the year ended. But one night in early February, 1968, he had the most unusual sighting of all. This was his account as carried in that same issue of CUFOR:

> "I was having a shower when it happened," he said. "We operate on diesel power at the ranch, so when the water suddenly stopped, I put on my dressing gown and slippers, grabbed a pocket flashlight and went out to the diesel shed to check the pump. It was just a small difficulty, so I was only in there a minute or two.
>
> "Now, the entrance of the shed looks out directly toward the corral where we keep the calves, and as I started to leave I noticed a dim light moving above the corral.
>
> "My small flashlight wasn't any help but there was just enough light to see what the thing looked like. It was disc-shaped, I'd say about 28 feet in diameter, with a dome on top and a dim light on top of that. I couldn't tell whether the rim was lighted or whether it was reflecting the light on top, but it was a little brighter than the rest of the object and seemed to be revolving. Around the base of the dome there was more faint light coming from what looked like three or four windows. I couldn't see anything inside."
>
> As the snow was melting in the unusually warm winter night, making a quagmire of the corrals, Grattan went back to the house to put on a pair of gumboots. Then entering the corral, he sloshed through the mud and stood almost directly under the object.
>
> "It was about 40 feet up, moving slowly along with a wobbly motion, and underneath it I could see three equally spaced markings pointing toward the center. They looked as if they had something to do with landing gear."
>
> Convinced after several sightings over his ranch that UFOs were not hostile, and in fact tended to shy away from humans, Grattan shone his small light at the object, waved his arms and tried as well as he could to show friendly interest.
>
> "It just didn't notice me at all," he said. "It just kept following the calves across the corral as if it was studying them, and all the time it was making a low rhythmic noise like an IBM computer."

Somewhere there may be a great corral in the sky where that study of animals continues. If that sounds too fanciful, let's remember that the still unsolved assault on livestock started with their unsolved disappearance in some cases. The Chomiaks in B.C.'s Playground of Gods were the losers in one such instance. Where do these animals go, since they do not seem to be chosen for surgery? For what may be an answer we have the information reportedly given by an alien in person, if we can call it that, during abduction.

Briefly, this case involved one John Higdon of Riverton, Wyoming. He said that while out hunting in October, 1974, he was confronted by an odd-looking stranger who somehow placed him, presumably mesmerized, inside a transparent "cubicle" along with five elk he had been tracking.

Higdon, accompanied by the stranger and another "man", was then transported to a distant point in space that his abductors called "home." However, it was somehow decided that Higdon did not qualify for whatever assignment the other two had in mind and, just as quickly, he was returned to earth. (APRO *Bulletin*, March 1975, & FSR, Vol. 21, nos. 3 & 4, 1975.)

Although the incident sounds imaginative, it has been taken seriously because Higdon passed an important test. In a state of hypnotic regression he was carefully questioned by Dr. R. Leo Sprinkle, who said, "Although the sighting of a UFO by a single witness often is difficult to evaluate, the indirect evidence supports the tentative conclusion that Carl Higdon is reporting sincerely the events which he experienced." It should be added that Dr. Sprinkle's inquiries by hypnosis into close-encounter cases like this and Judy Doraty case have proved a notable contribution to study of the UFO mystery.

The point in mentioning this one is that while under hypnosis Higdon recalled one of his abductors as saying that animals were carried off to their home planet for breeding purposes.

As might be expected of our visitors, that rarely given insight into their motives raises more questions than it answers. For instance, if they are producing their own animals up there, why do they need parts of ours?

But that sort of questioning will get us nowhere. It typifies the human assumption that if we cannot understand something, it contains no sense. When the day of the great close encounter comes, let's hope we have progressed from that.

As it happens, we are receiving some useful training for that time in our ever-expanding study of the stars. Now that we are learning to accept the impossibilities of space, we may have reached some maturity about such matters when we meet its people who otherwise might also seem "impossible."

Also, although even some UFO-watchers will question the point, we are already progressing toward that time because of what we learn about our UFO visitors. Unspoken or otherwise, and all the different theories notwithstanding, what we really look for in ufology is a picture of what awaits us when we encounter another civilization in space, and by minute degrees a view is now taking shape.

So, in hope that these pages will help in adding some small bit to what we can see, let's re-cap briefly before we conclude.

We have had a look first of all at the concern that the aliens do seem to show about the affairs of this planet.

More than anything, this seems to confirm our starting premise that the visitors generally approach earthly problems with care and respect.

Then we have seen that their use of bell-jar magic does indeed justify our giving them credit for skills far beyond ours. Yet we have observed their continuing search for electrical power, which seems to expose a weakness in their armature that makes them almost like our own kind.

Finally, we have studied the aliens' invasion of cattle ranches, and here it is hard to give them credit for "care and respect" in their approach to earthly matters. At the same time we have noted in an eye-witness case that they perform their surgery with a business-like air as if it was for "our betterment." While it is therefore too soon to give a verdict on that issue, we have learned a great deal more about our visitors' methods and personality.

So as we come to know these strangers much better — even though any real understanding may take thousands of more years — and when we eventually reach that remote time when the relationship is fully acknowledged at some distant point in space, or perhaps on earth, where do we go from there? Will that mean the end of the human story as it is now spelt out in terms of life on an isolated planet?

Almost certainly it will. Once we have joined the space fraternity we can never retreat from it any more than primitive tribes on earth can regain their privacy once they have lost it. By that same token, however, it will be far from the end of the total human story. If our crystal ball is clear enough to take us far into the future, what we dimly see might be the time when we live with our space friends on another world and become absorbed by them.

Perhaps that happens when, under a bloated dying sun, we flee earth and then the solar system to escape the killing heat. To recall what was said in chapter two there seems little doubt that earlier civilizations in space, threatened with utter destruction of some sort, have been through similar ordeals. Maybe some of those refugees became part of our own ancestry.

Beyond that period of exodus our crystal ball becomes hopelessly dim. Yet those with special gifts of observation may still detect some slight movement in its clouded center. It then comes to mind that maybe no life of any species is ever lost, but only recycled, and with this we are reminded again of Antonio's ordeal.

Perhaps through the telescopic medium of an alien bell-jar, with screening equipment added, he did indeed look beyond the ultimate ET experience. Possibly he peered into a realm where time appears as a dimension like space in providing the same freedom of movement instead of simply one course forward. Conceivably that would mean one could travel in various directions of time and reach any destination instantly since time within itself, like a drop in the ocean, would not be a separate factor.

While that would not explain why Antonio saw himself on the aliens' screen, it could explain why anyone thrust into that position would suffer loss of memory. The sense of disorientation in such a world with its kaleidoscope of changing views as the victim is moved instantly from scene to scene (although to Antonio the pace at each stop somehow appeared normal) would be overwhelming.

But we must wait until the aliens let us use their bell-jars freely before we can systematically examine what the future holds. Meanwhile there is much to keep us busy in probing the history of the universe with our own instruments. As we listen to the fading whispers of the "big bang" and extend our sight to the outermost fringe of stars, we begin to have some picture of our field of operations.

If our rate of discoveries continues, we should just about finish our assignment with the past before any bell-jar work on the future begins.

Acknowledgments

Material in this book borrowed from the following publications and services, in quoted form or by reference, is gratefully acknowledged.

Aerial Phenomenon Clipping & Information Center, P.O. Box 9073, Cleveland, Ohio 44137, U.S.A.

APRO Bulletin; Aerial Phenomenon Research Organization, Inc., 3910 East Kleindale Road, Tucson, Arizona 85712, U.S.A.

CUFOS; Center for UFO Studies, P.O. Box 1621, Lima, Ohio 45802, U.S.A.

Flying Saucer Review; FSR Publications Ltd., West Malling, Maidstone, Kent ME19 6JZ, England.

Frontiers of Science, 2201 St. Paul Street, Baltimore, MD 2128, U.S.A.

Harmonic 33 by Bruce Cathie. A.H. & A.W. Reed Ltd., 65-67 Taranaki St., Wellington, N.Z.

Inforespace, Avenue Paul Janson, 74 - 1070 Brussels, Belgium.

Journal UFO (discontinued), U.P. Investigations Research Inc., P.O. Box 455, Streetsville, Mississauga, Ont., Canada L5M 2B9.

Le Proces des Soucoupes Volantes by Claude MacDuff. Les Editions Quebec-Amerique, Case Postale 96, Montreal, P.Q., Canada H1X 3B6.

MUFON UFO Journal; Mutual UFO Network, Inc., 103 Oldtowne Rd., Seguin, Texas 78155, USA.

National Enquirer, Lantana, Florida 33464, U.S.A.

Our Mysterious Spaceship Moon by Don Wilson (1975), and *Secrets of Our Spaceship Moon* by Don Wilson (1979), Dell Publishing Co., Inc., 1 Dag Hammarskjold Plaza, New York, N.Y. 10019, U.S.A.

Overlords and Olympians by Wm. Gordon Allen. Health Research, P.O. Box 70, Mokelumne Hill, Ca. 95245, U.S.A.

Stigmata, Project Stigma, P.O. Box 1094, Paris, Texas 75460, U.S.A.

The Mysterious Signals from Outer Space by Duncan Lunan. Bantam Books, Inc., 666 Fifth Ave., New York, N.Y. 10019, U.S.A.

The Night Mutilators by Gene Duplantier. SS&S Publications, 17 Shetland St., City of North York, Willowdale, Ont. Canada M2M 1X5.

The 12th Planet by Zecharia Sitchin. Avon Books, 959 Eighth Ave., New York, N.Y. 10019, U.S.A.

The UFO Connection by Arthur Bray. Jupiter Publishing, P.O. Box 5528, Station F, Ottawa, Ont. Canada K2C 3M1.

UFO Newsclipping Service, Route 1 - Box 220, Plumerville, Arkansas 72127, U.S.A.

UFO-Quebec, Case Postale 53, Dollard-des-Ormeaux, P.Q. Canada H9G 2H5.

UFO-SVERIGE, Box 311, 591 03 Motala, Sweden.

U.S. Air Force Project Blue Book, Special Report No. 14, Saucerian Books, Box 2228, Clarksburg, W. Va. 26301, U.S.A.

A Strange Harvest (film) by Linda Moulton Howe. KMGH-TV7, McGraw-Hill Broadcasting Company, Inc., 123 Speer Boulevard, P.O. Box 5007, Denver, Colorado 80217, U.S.A.

In addition, special thanks to artist Brian James whose generous work helps so much to illustrate the action of *Aliens Above*; to correspondent Graham Conway for his interest and alert reporting; and to Beatrice Zimmer for her fine study of abductions, held for later use.

Copies of this book at $11.50 each, Canadian or U.S. (postage included) and signed if you wish, may be ordered direct from the author. Outside North America add $1.00. Write:

John Magor
P.O. Box 758,
Duncan, B.C.,
Canada V9L 3Y1

John Magor, a professional newspaperman, is the former editor and publisher of the widely-read *Canadian UFO Report*. He holds a Master of Science degree in journalism from Columbia University and he was an RCAF coastal-command pilot during World War II. This book is his second on the UFO mystery. His first book, entitled *Our UFO Visitors*, was published in 1977 by Hancock House.

Author's Note

Originally, the title for this book was simply *Aliens Above*. I thought nothing more was needed to convey the picture of a strange race surveying our planet.

However, close to press time three separate, but seemingly related, incidents happened almost together. First, the editor at Hancock House phoned to ask if I would write a dedication for the book. Second, word came in the mail that *Frontiers of Science*, a fine magazine which was closely concerned with the UFO subject, had folded. Third, I came across a previously overlooked news story with the heading, " 'No one pays attention' to UFO reports any more." (Chicago *Sun Times*, Nov. 4, 1982)

The last item was based on an interview with Dr. J. Allen Hynek, the head of CUFOS and a crusader for a more serious attitude toward UFOs. He observed that although his Center received as many remarkable sighting reports as ever, people now said, "So what else is new?"

Dr. Hynek had, of course, correctly stated the case. To those of us in the UFO reporting business, the decline of interest in UFO activity is apparent. Trouble such as the folding of *Frontiers of Science* seems to be one result of it.

However, we who write about these strange visitors must take part of the blame. Since readers often want what is bizarre and exciting, that is what we are apt to serve them until it becomes tiresome. Instead we should look more for the purpose of the aliens because it is there we will find their real identity and there, too, must lie the most fascinating part of the story. To practise what I preach, I have tried to do that in these pages even though it means drawing on some incidents already well reported.

Still, the bottom line remains — our UFO visitors have been here a long time and obviously intend to stay, whatever we say about them. At some time in ages to come we may learn to communicate and perhaps even to live with them, as speculated in my conclusion. Who then will ask what else is new?

That thought should have eased my press time doubts which the second and third incidents aroused. But believing there is some still undefined law of coincidence, I suspected the simultaneous occurrence of these three incidents might be a message for me to do something and I decided to act on the first incident.

I then added "always" to the title and so dedicated this book to the enduring nature of the UFO presence.

INDEX